ABOUT THE AUTHOR

Robert Tiley's interest in antique books and maps, the Australian out-doors and a curiosity for things historical led him to unravel the personal and political drivers of early Australian maritime exploration. A lawyer by training and a financier by trade, Robert worked in London, Singapore, Beijing and Tokyo during the 1980s before return-ing to appreciate the unique charms of Australia.

Robert lives in Sydney with his wife Jean, and children Sam and Lucy. They patiently tolerate his other passions of rowing, restoring wooden boats and fiddling with old cars.

Robert Tiley

Australian Navigators

PICKING UP SHELLS AND CATCHING BUTTERFLIES IN AN AGE OF REVOLUTION

Kangaroo Press

AUSTRALIAN NAVIGATORS
First published in Australia in 2002 by Kangaroo Press
An imprint of Simon & Schuster (Australia) Pty Limited
20 Barcoo Street, East Roseville NSW 2069

A Viacom Company
Sydney New York London

Cataloguing-in-Publication data:

Tiley, Robert, 1959- .
Australian navigators : picking up shells and catching
butterflies in an age of revolution.

Bibliography.
Includes index.
ISBN 0 7318 1118 6.

1. Explorers - Australia. 2. Australia - History - To 1788.
3. Australia - Discovery and exploration. I. Title.

994.01

Cover and internal design by Gayna Murphy, Greendot Design Pty Ltd
Typeset in Adobe Garamond 12 point on 15 point leading
Printed in Australia by Griffin Press

10 9 8 7 6 5 4 3 2 1

For Jean, Sam and Lucy the Boosterfish

ACKNOWLEDGMENTS

With thanks to Geoffrey Ingleton, Frank Horner, Patricia Cornell and the others who have brought so much of the navigators' work to the public.

To Rose Creswell and Annette Hughes for their enthusiasm and encouragement.

To Angelo Loukakis, Brigitta Doyle and Anne Reilly for their advice and guidance.

To Jean, Sam and Lucy for their love, patience and support.

CONTENTS

Preface *ix*

CHAPTER ONE Introduction *1*

CHAPTER TWO The first flashpoint, the Treaty of Paris *6*

CHAPTER THREE *Endeavour*, a groundbreaker for the British *15*

CHAPTER FOUR An 'Enlightened' and ascendant Banks *21*

CHAPTER FIVE Cook's school *26*

CHAPTER SIX Disaster for Britain, opportunity for Banks *31*

CHAPTER SEVEN Essentials for the navigators' expeditions *37*

CHAPTER EIGHT William Bligh, the survivor of Cook's school *53*

CHAPTER NINE The new blood, Matthew Flinders *64*

CHAPTER TEN Flinders' bold dash *79*

CHAPTER ELEVEN Little *Lady Nelson*, the forgotten
discoverer of Port Phillip *85*

CHAPTER TWELVE The French reappear *90*

CHAPTER THIRTEEN D'Entrecasteaux's rescue mission *98*

CHAPTER FOURTEEN Baudin, the first big French success *106*

CHAPTER FIFTEEN Flinders' blunders, and tragedy *123*

CHAPTER SIXTEEN Flinders and Baudin – the race begins *130*

CHAPTER SEVENTEEN Flinders heads north *143*

CHAPTER EIGHTEEN Baudin heads south *150*

CHAPTER NINETEEN Flinders' luck runs out *163*

CHAPTER TWENTY Two replacement ships *170*

CHAPTER TWENTY-ONE Mauritius, a powderkeg for Flinders *181*

CHAPTER TWENTY-TWO Flinders' nemesis, Decaen *185*

CHAPTER TWENTY-THREE The race for publication *192*

CHAPTER TWENTY-FOUR The French publish the first
map of Australia's coastline *197*

CHAPTER TWENTY-FIVE Flinders published at last *204*

CHAPTER TWENTY-SIX The achievements *211*

CHAPTER TWENTY-SEVEN Conclusions *214*

Appendices *220*

Endnotes *228*

Glossary *233*

Suggested further reading *237*

Index of People, Places and Ships *241*

PREFACE

One Sunday in 1999 I wandered curiously into the sales rooms of a large Sydney auction house. One of the final auctions of historian Geoffrey Ingleton's book and map collections was about to happen. I knew nothing of Ingleton and had the usual Anglo- (and slightly Dutch-) oriented knowledge of Australian maritime history that you would expect from a kid educated in the early 1970s ... all Tasman, Dampier, Cook, Bass and Flinders. I couldn't recall being taught much about Bougainville or Baudin or even hearing names like Grant, Thistle, Freycinet or the *Lady Nelson*.

I realised from leafing through some of Ingleton's books that there was so much more. Name after name turned up of individuals involved in Australia's early coastal exploration, all of whom had struggled to come to terms with this very foreign environment. Even conveying this in pictures was difficult. The published narratives of Bougainville and Cook, no doubt embellished by their publishers, portrayed Tahitians in a classical Greek or Italian vein; pictures of the ubiquitous kangaroo meanwhile looked far more deer- or rabbit-like. Though they would never have seen the like before, the artists still portrayed their subjects in terms far more familiar to a European eye and misleading for that.

Among Ingleton's maps was a beautifully illustrated slim little atlas. Within its fourteen plates was the first complete map of Australia ever published. While I didn't realise the importance of this map at the time, my curiosity was aroused. The atlas was French, not British, and I'd never even heard of its draughtsman.

From then on I was hooked. I wanted to understand why so many maritime explorers had come to – or near – such a foreign and potentially dangerous place, mostly in the last forty years of the 1700s.

The Dutch had been there long before, largely by accident – in most cases literally so, following shipwreck. The great Dutch navigator Abel Tasman was a notable exception. Tasman and other early explorers, such as the lesser-known de Vlamingh, while undeniably Australian navigators, are not the focus of this book. Although they charted almost three-quarters of the Australian coastline, the Dutch were ultimately only concerned with keeping well away from its dangerous, infertile shorelines and its inhabitants, and they didn't return.

Yet the likes of Bougainville, Cook, Bligh and Baudin, those asked to find and chart new coastlines in the Pacific over this period, were required to do much more, to engage in an incredibly hazardous task. They were expected to take their ships into this uncharted domain and accurately record its coastlines and hazards, and also note useful information for trade or science. Their ships were complicated to manoeuvre, and couldn't simply be thrown into reverse with the flick of a switch at the first sign of danger. They required, like all machinery, constant care and attention to survive the long voyages, as did their crew. Mortality on these sea journeys was high. Few of the French navigators survived their first major expedition to the Pacific, and very few of the British commanded more than one. Only one, William Bligh, ever achieved what could be regarded as comfortable retirement.

A great deal has been written about many of these navigators, often in weighty volumes. Without the scholarly works of the likes of Scott, Lee, Mackaness, Beaglehole, Ingleton, Cornell and Horner, my research would have been enormously more difficult. This book isn't trying to emulate such efforts. Instead it seeks to address the spirit of the era that inspired their writing, to examine afresh the personal ambitions of many of the navigators, the quite often different goals of their political patrons and backers, and the reasons why they did or didn't succeed.

Although it was their skill, humanity and courage that would deliver the navigators from the dangers of coral reefs at sea or cannibals ashore, they were as much pawns in the floating political world of the late 1700s as many a corporate executive may find him or herself today. They are fascinating studies of human tenacity and frailty. They are as relevant now as then.

CHAPTER ONE

INTRODUCTION

On 13 September 1759 a young French colonel was desperately urging 2000 experienced troops along a small road in North America. He was aware that he was running out of time. He was racing to reinforce a ragtag French army, largely composed of local undisciplined militia, defending Quebec from a seasoned force of British regulars that had surprised them with a virtually unopposed landing above the town early that morning.

He would arrive only to witness a French defeat; his commander already lay among the fallen. By then the British had decimated the French after a savage campaign in a battle that even the British commander, General James Wolfe, hadn't believed he would win or, personally, survive. He was partly right, as he also lay with the dead. Quebec would surrender shortly afterwards. The British victory was tenuous, with winter approaching, their supply lines stretched and with a French counterattack already in process. But the British had achieved their objective. They had moved an army by ship up the St Lawrence River and had finally taken Quebec.

On the river below, a British ship's master of about the same age as the young colonel was at last soundly asleep after months of dogged work marking channels, manoeuvring his captain's ship in the current and surviving attacks from the Indians, the French and their fire ships in order to deliver the British troops to battle. Europe was still in the grip of the same war, which would terminate with the Treaty of Paris after seven long years.

The French colonel and the British ship's master would never meet. They couldn't begin to imagine then that although the battle and the peace that ultimately followed would each have its place in history, they themselves would figure far more prominently in history half a world away in Australia, in many ways as a consequence of that battle, the campaign which led to it and the events that followed. They were Louis Antoine de Bougainville and James Cook.

Accelerating interest in the Pacific

Australia lay for centuries like an unread book with a boring cover. The early Dutch explorers to have contact with its western coastline recorded that they saw nothing more notable than sand and shipwrecks, and didn't examine it further. A succession of shipwrecks on the west Australia coastline, from *Batavia* in 1629 to *Zuytdorp* in 1712 and *Zeewijk* in 1727, left an historical testament to the dangers here.

One of the most audacious of the Dutch explorers was Abel Tasman. He established that the Australian continent was separate from New Zealand and the Antarctic, and mapped much of its northern coastline. But even this could excite little interest – notwithstanding the proximity of the Dutch bases in Java to Australia, the support of his strong patron, Governor Van Diemen, and the relative ease of access the Dutch had when compared to later European navigators. The continent appeared to be barren, devoid of any commercial value, and peopled by an extremely hostile race. After Tasman's departure the continent lay unvisited for the best part of 130 years, apart from fleeting interest by the likes of the Dutchman de Vlamingh and the English adventurer William Dampier at the end of the seventeenth century.

But by the end of the eighteenth century things had dramatically changed. Rival expeditions – the best equipped to date – from France and Britain, who

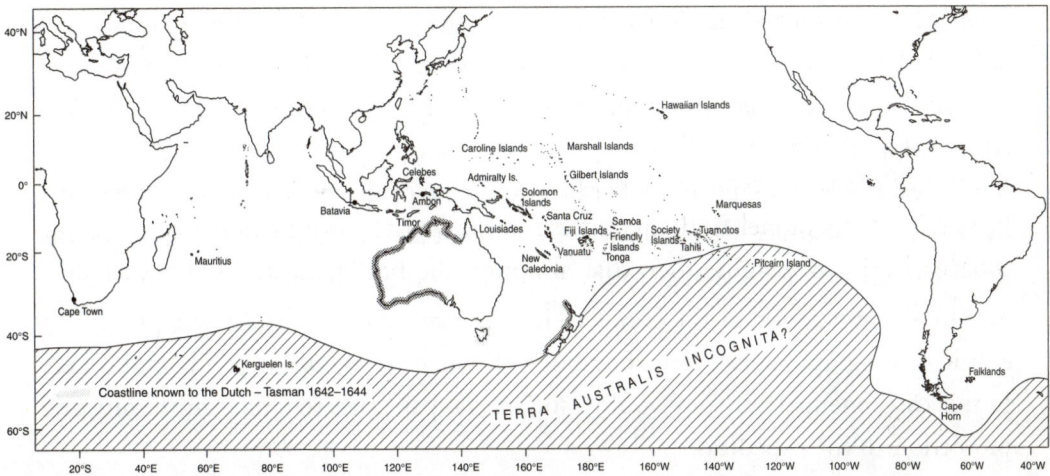

Modern-day names, Dutch discoveries in Australia and the suspected location of *Terra Australis Incognita* pre-1767

were at war with each other again, were examining Australia's coastlines. By early 1802 there were no less than four survey vessels at different points on the Australian coastline.

Why had interest accelerated in this continent so many years after Tasman's visits? Why was exploring the coastlines attractive to battling nations half a world away when their resources were stretched thin funding war? And why would sailors opt for tedious, dangerous and unrewarding survey work when they could accumulate so much more position, power and prize money by fighting the wars? When so few of the significant navigators of Australian coastlines ever won wealth, position or longevity, despite their apparent success, it's difficult to grasp what spurred them on.

The answers lie generally in the political and scientific developments of the period and, more specifically, in the qualities a navigator required to successfully survive in the political chaos of late eighteenth-century Europe.

The navigators' voyages over the period of 1763–1803 were often performed for disparate reasons, without overarching strategy. This was a period of considerable European turmoil, with the equivalent of three world wars and between them the

French Revolution and the American War of Independence following in quick succession. These major flashpoints disrupted international relations and trade and produced massive social unrest and insecurity. The navigators' voyages were often ad hoc responses to these flashpoints rather than purposeful, long-term strategic initiatives. It was difficult to predict how long any of these conflicts – or the short-lived periods of peace that followed each – would last. Ultimately, whatever the underlying reason for a voyage to the Pacific, it could only proceed without fear of capture or destruction at the hands of its adversaries if its purpose was purely scientific or geographic discovery, the results of which would ultimately be published. In this case, enemy countries would issue special passports to confer immunity and protection to the expedition.

For the individuals who might lead such voyages there were competing attractions. The political tumult of the era could create heroes and fortunes among the multitudes of naval officers in the hugely competitive and nepotistic world of the eighteenth century where connections, or 'interest', meant everything. War was what most yearned for. In wartime and with luck and courage, a well-connected officer who had managed to obtain command of, say, a fast frigate could make a fortune in prize money. In peacetime he was then less likely to find himself ashore, making ends meet on half-pay without a command. He wouldn't have been immediately attracted to becoming a navigator. Such an impoverished existence was the usual prospect for the less well-connected naval officers, which was what most of the British and French navigators initially were. Exploratory work, although so much longer and more exposed to the dangers of death from disease or shipwreck – and therefore more dangerous than normal naval service – provided an escape from this dismal prospect, as well as a chance for a degree of fame and glory from new discoveries and royalties from publication of their exploits.

To achieve fame and fortune a navigator required a range of skills. He needed first of all to achieve through a mixture of knowledge and nepotism sufficient rank to be noticed. He had to maintain control over poorly fed men engaged for years in a dangerous task in confined, unhealthy conditions. He required exceptional know-how, courage and nerve as a seaman to find his way through uncharted waters. He needed sound mathematical and drafting skills, coupled with a basic knowledge of astronomy to accurately record the path he had taken, as well as the dangers, for those who would follow, and equally as accurately to commit this to paper. Crucially, he needed onshore patronage and financial backing from government and scientific institutions to promote, properly provision and execute the voyage, to provide him with a measure of immunity from enemy attack in the form of diplomatic passports, and to publish the expedition's results. Finally, he needed political savvy with native populations abroad and with political change at home.

Over all this, he needed a bucketful of luck.

Of the many navigators, only a handful actually survived to enjoy any fruits of their labours. One was William Bligh, notorious as a tyrant but misrepresented by history. Luck played little part for him, but he was about the only British navigator to reach what might be termed comfortable retirement. One who didn't was Matthew Flinders, who was in many ways far luckier. Despite Flinders' labours, however, it would be a rival French expedition that was to publish the first complete map of the Australian continent, in an unimposing little atlas. Tensions were so rife within the French expedition that the name of its leader, Nicolas Baudin, was nowhere mentioned in its official narrative or its published maps. Baudin wasn't the only French navigator to suffer.

As the interest in the Pacific developed, the British, by chance, gained a significant advantage over all other nations, particularly the French. It had happened through a combination of circumstances that took two remarkable men from almost opposite backgrounds on an even more remarkable voyage. They were James Cook, that young master from the Quebec campaign, now an ageing junior officer in the navy, and Joseph Banks, a young and wealthy landowner. The voyage was of *Endeavour* in 1768.

That the voyage had exceptional consequences for the British is well known. That many of these consequences were unintended and unforeseen at the time; that they had been made possible only because of some unique ingredients of the voyage in particular and some fortunate historical flashpoints in general; and that the French were often further ahead of the game – these things are not so well known.

CHAPTER TWO

THE FIRST FLASHPOINT, THE
TREATY OF PARIS

The first event to revive the interest of Western powers in the Pacific was the end of the Seven Years' War. This had been the equivalent of a world war at the time. It commenced in 1756, and was fought in Europe, India and North America. It finished disastrously for the French at the Treaty of Paris in 1763, one of two comprehensive treaties required to settle the large number of differences between almost all European countries. The surrender of the French to the British at Quebec had marked the beginning of the end of the North American war. Under the Treaty of Paris, France renounced to Britain rights to almost all its North American colonies east of the Mississippi, some of its West Indies settlements and all its gains in India and the East Indies dating from 1749.

France was stripped of almost all its colonial and global ambitions in America and India, where British expansion was most active. The treaty created for France a massive desire to rebuild, just as the onerous terms of the Treaty of Versailles would do for Germany after World War I.

The French reaction

Louis Antoine de Bougainville never had the chance to counterattack and throw the British out of Quebec. He had to contend with winter and stretched supplies while the British received fresh supplies and reinforcements. In the space of two years Bougainville took part in the French withdrawal following their surrender in North America in 1760, was later captured by the British then released again, only to be wounded in yet another action, in Germany. With the Treaty of Paris pending, the French Army had little need of his service and he looked for a different vocation. He proposed to the French Government a project intended to rebuild French trade and commerce to the Pacific and Asia. His plan was to settle a French colony in the Falkland Islands. The settlers were displaced North American French ex-colonists forced out by the Seven Years' War. The government agreed to the project, although Bougainville and other individuals had to privately finance it.

The Falklands were more than just a dumping ground for homeless colonists. Bougainville, and through him the French Government, saw these islands as strategic to the control of the eastern entrance to the Strait of Magellan and to Cape Horn, the only western routes for trade between Europe and the Pacific. Here France could rebuild its overseas trading bases, as the British and Dutch denied access to the Orient and the Pacific via the only other route to Asia, the eastern route through the Cape of Good Hope and the East Indies. Bougainville duly located the Falklands and settled his fledgling colony in the eastern group in 1763.

Spain objected to Bougainville's occupation, however, as likely to re-awaken long dormant interest on the part of their British enemies in the Falklands for similar strategic reasons. In order to maintain relations with their Spanish allies, who agreed to compensate Bougainville and his other backers for their financial loss, the French Government ordered Bougainville first to remove the colonists from the Falklands, and then to explore the Pacific.

The Spanish claim was based on a centuries-old treaty with the Portuguese, the Treaty of Tordesillas of 1494, which had effectively divided the earth between the two dominant powers at the time, Portugal and Spain. Although the world order had long changed, political tensions after the Seven Years' War were such that France saw no reason to anger its ally over these islands.

Once his Falklands colony had been dismantled, Bougainville's new orders were to ascertain whether or not the unknown southern continent, *Terra Australis Incognita*, existed, and to seek new sources of trade. The Treaty of Paris had also destroyed the French India Company's century-old monopoly over trade in the region, allowing enterprising French nationals such as Bougainville to exploit this new freedom. Therefore Bougainville was also interested in finding another sea route to the Orient, and possible bases to support this, to replace the eastern French trade routes relinquished at the Treaty of Paris.

Terra Australis Incognita wasn't Australia. Australia itself was nothing new or unheard of. It was then known as New Holland, and had been since the Dutch visits in the early 1600s. It was just not fully defined, with only its western, south-western and northern coastlines sketched out. It was certainly sufficiently defined that everyone knew it wasn't the fabled *Terra Australis Incognita*. Tasman had proved this in 1642.

Tasman had also seen the western shores of New Zealand. It was generally believed that this shoreline was probably the western coast of a large landmass extending further east across the Pacific to Cape Horn, and that New Zealand was merely its western boundary. Bougainville was now asked to find it.

After decommissioning the Falklands colony, Bougainville headed for the Pacific in November 1767. Voyaging until 1769, Bougainville's expedition would encounter Tahiti and its delights and rediscover parts of the Solomon Islands and the New Hebrides, not seen since the Spanish had visited almost two centuries beforehand. In seeking *Terra Australis Incognita* in the Pacific, Bougainville was the first explorer since Torres, 160 years earlier, courageous enough to attempt to penetrate the mysteries of what lay towards the Australian Barrier Reef. In early June 1768 he turned away just off the Queensland coast as he approached a long stretch of breakers. Bougainville Reef, east of Cooktown and some 100 kilometres from the Barrier Reef, now marks this spot.

Bougainville's decision to turn away from the breakers was probably sensible, given a hungry and weakening crew, increasing fields of shallow coral surrounding him and adverse winds cutting off any escape. As he sailed north through what

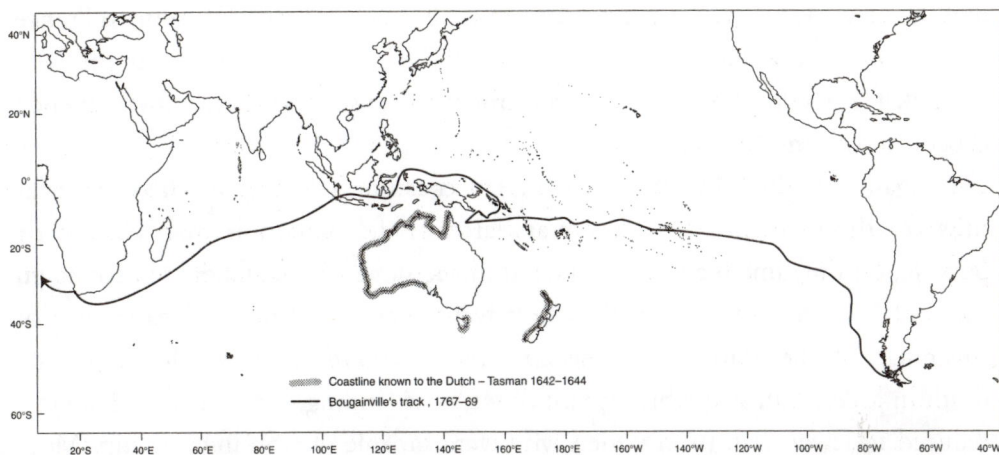

Bougainville's near miss, June 1768

he named the Louisiades, his men's hunger was such that they were trying to eat the leather from the yards. He sailed on through the Solomons, past Bougainville Island and the north coast of New Guinea, through the Moluccas (near Sulawesi) to Dutch Batavia (Java). Seeing the trading bases of the Dutch, Bougainville became aware for the first time of the potential richness of these islands. He was also conscious of likely British competition in the area.

Despite his near miss with the Australian coastline, his voyage was regarded as a huge success in Paris. He brought back from Tahiti a live Tahitian, Ahu-toru, over whom the poets and philosophers in the salons of Paris went wild. Here was personification indeed of one of the subjects of Jean-Jacques Rousseau's *Discourses*. Rousseau had written these in the 1750s, claiming that man, though born neither good nor bad, was incapable of finding happiness due to the excessive complexities and inequalities of wealth in French society. True happiness could only come, he argued, in a community small enough to be understandable to man and one in which he could have a real part in its government. Ahu-toru seemed such a man, and Tahiti such a place.

Bougainville published a narrative of his voyage in 1771, and again in English in 1772, reaching a wide public. From over two years of sailing, his two ships had lost only nine men from illness. The French had relocated parts of the Solomons

9

and the New Hebrides, discovered the Louisiades, and realised the value to France of further exploration in the area. Bougainville's own position from this was assured. He would have a major involvement in most subsequent official French expeditions to the Pacific.

Bougainville's brush with the Australian coastline wouldn't be the only 'near miss' for the French. Only a few years later, his fellow countryman Charles de Surville, on the same New Zealand coast at the same time as Cook, elected to go east rather than west like Cook. Cook went on to delineate the east coast of Australia. Surville continued east seeking *Terra Australis Incognita*. He found only death in a Peruvian surf while attempting a boat landing, his scurvy-ridden crew reduced to fewer than twenty men who were capable of working his ship. More 'near misses' by the French would follow.

British interest

Although Bougainville's endeavours in the Pacific were arguably more successful than those of any other European ship to date, in fact four British ships had recently preceded him across the Pacific in the mid-1760s. This was initially more by circumstance than design, but the results would trigger Anglo–French rivalry in this region. This sequence would be repeated in the following sixty years with La Pérouse and the First Fleet in 1788, Baudin and Flinders in 1801, and Freycinet and King in 1818.

Independently of the French, the British were also pursuing an interest in new trade routes around the American continent to the East Indies and China via either a northern or southern route. In the case of the latter, through the Strait of Magellan, the British had as early as 1748 considered settling a colony in the Falklands but hadn't proceeded due to the likely Spanish objection.

In 1764 and in closely guarded secrecy to avoid antagonising the Spanish obsession with the Treaty of Tordesillas, Commodore John 'foul weather Jack' Byron – grandfather of the poet – had been sent in a frigate, *Dolphin*, with the smaller *Tamar*, to discover 'Pepys Island', rumoured to exist in the South Atlantic. He was then to sail up the west coast of America to find a passage across its north, the 'North-West Passage', and sail back to England, effectively achieving for Britain what Bougainville was attempting to achieve for France.

Byron hardly adhered to the impossible instructions. He didn't find Pepys Island but did relocate the West Falkland Islands. They had the same strategic value to Britain as they did to Bougainville. In fact Bougainville was already settled in the East Falklands with his colonists when Byron visited the western group, which Byron didn't realise at the time. He discovered their existence not long after his departure and immediately sent word of the discovery to Britain.

Byron then completed the circumnavigation through the Strait of Magellan, into and across the Pacific – by almost the most direct and fastest route possible – and home by the Cape of Good Hope without adding much by way of new discovery at all. Although some aspects of *Dolphin*'s seaworthiness may have constrained Byron, it couldn't be said that he had Bougainville's mental toughness or curiosity.

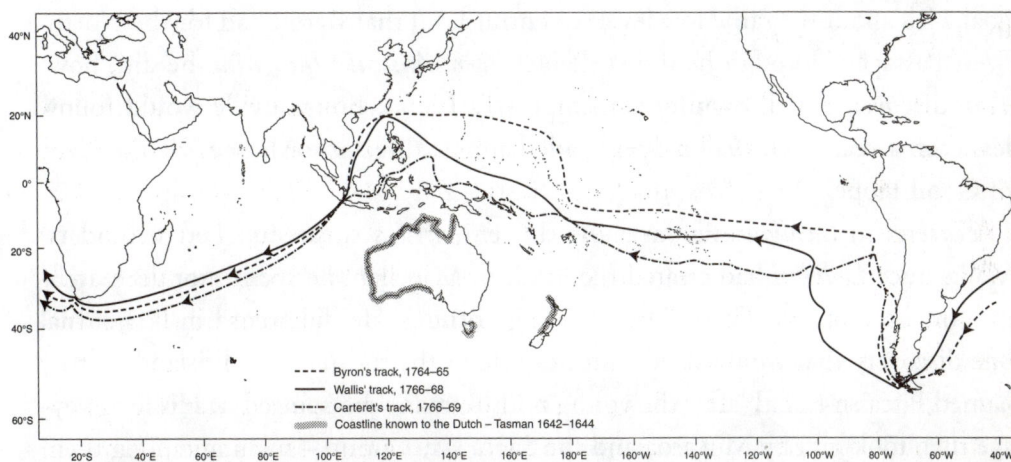

Tracks of Byron, Wallis and Carteret in the Pacific

News of Bougainville's settlement at the Falklands galvanised immediate action by the British, just as the Spanish had feared, and a small British garrison was landed at Port Egmont in the West Falklands in early 1766. Despite strong Spanish objections, culminating in an attack on the settlement, almost ten years would pass before its removal.

Disappointed by the paucity of results from Byron's voyage, the British Admiralty sent out Captain Samuel Wallis shortly after, in July 1766, again in *Dolphin*, with many of Byron's crew and with orders to find *Terra Australis Incognita*. This was also believed to be Davis Land, land west of Peru. In *Batchelor's Delight* in 1687, the English buccaneer Edward Davis claimed to have discovered it, and fellow buccaneer William Dampier also mentioned it in his writings. Again the British and the French were chasing the same quarry. Bougainville, still in the process of dismantling the French Falklands colony, now had similar orders to Wallis, for similar reasons, namely pursuit of a new trade route to the East. With Wallis went Philip Carteret, in the small, slow-sailing and poorly supplied *Swallow*, plus a store ship. Carteret had sailed with Byron and had some idea of what he was in for, particularly when told that most of his small ship's supplies would be carried by *Dolphin*. Wallis looked for his goal with about the same low levels of enthusiasm that Byron had for the North-West Passage. Although he didn't deliver *Terra Australis Incognita*, he did, however, discover Tahiti, naming it George III's Island. Bougainville would follow less than a year after. Wallis then headed almost directly for home, via the Cape of Good Hope.

Carteret in *Swallow* was far more diligent, if less successful. Left behind by Wallis once *Dolphin* had cleared the Strait of Magellan, he spent months searching for the supposed Davis Land, finding nothing. He did record in his journal one discovery that would have a greater role in the future: a small island that he named Pitcairn Island, after the young midshipman who spotted it. His long voyage then took him by Mururoa and the Santa Cruz group – naming among them Ourry's Island, north of Vanikoro – then on to New Britain and New Ireland, north of New Guinea, to a group he named the Admiralty Islands, which would later distract d'Entrecasteaux from his goal.

As Wallis returned to Britain in May 1768, Carteret was still dodging Dutch obstinacy, bureaucracy and intrigue to get *Swallow* to Batavia through Makassar Strait for a refit, and then home via the Cape of Good Hope. On 19 February 1769, on the home stretch, he was passed by the now homeward-bound Bougainville, who caught up with him and attempted to extract from him as many details of his voyage as possible, while not passing on any of his own.

Carteret reached Spithead on 20 March 1769, almost a year after *Dolphin*. By this time *Endeavour* had been at sea for eight months.

Despite their scant results, the voyages of Byron and Wallis are most relevant because they used largely the same crew, in *Dolphin*. Already the British were building a stock of experienced long-distance sailors. This would make all the difference to the success of James Cook, whose courage and audacity at least equalled Bougainville's and Carteret's. Cook would get the benefit of many of *Dolphin*'s crew for *Endeavour*, particularly some of his officers, including in most of his future voyages senior officers John Gore and Charles Clerke.

James Cook, the surveyor's apprentice

The Seven Years' War and British consolidation of new possessions after the Treaty of Paris ignited French interest in the Pacific. Coincidentally, these circumstances provided James Cook with the opportunity to train in basic coastal hydrography, namely coastal marine surveying and shallow water navigation. This would prove to be integral to his exploratory work in the Pacific.

Cook's superiors had noted his abilities when he assisted in navigating the long St Lawrence River before the taking of Quebec. Cook, an experienced sailor from commanding colliers and other trading vessels in the North Sea, had volunteered for the navy at the relatively late age of twenty-six. With his experience he had risen quickly to non-commissioned rank, such that during the St Lawrence campaign, as a ship's master, he was responsible with others for buoying the channels for the British invasion fleet.

During the campaign an experienced engineer he had befriended, Samuel Holland, introduced him to the science of surveying. Holland had been commissioned to chart the nearby bays and estuaries. With the support of Cook's influential commander at the time, Captain John Simcoe, Cook became closely involved in the survey and would have learnt much from Holland.

Cook was already choosing his contacts well. Samuel Holland would become the surveyor-general for Newfoundland.

Following the Treaty of Paris, comprehensive charts were required of the new coastlines acquired from France. This included the St Lawrence River and coastlines and areas of Newfoundland and Nova Scotia. Cook's newly acquired chart-ing skills and connections through Holland and Simcoe, and other commanders Lord Colville and Palliser, placed him perfectly for the job and he spent the next five years or so producing charts of these areas most efficiently. The experience Cook acquired as a result, and the attention his charts received from the Admiralty, would serve him well in the future. During this time he'd also attracted the atten-tion of the Royal Society after noting and measuring an eclipse of the sun in Newfoundland in 1766 and sending it the results.

The Royal Society was a powerful body, formed in 1660 very much on the basis of Francis Bacon's 1620 maxim that 'human knowledge and human power meet in one'. The Royal Society was comprised of unpaid – and therefore usually wealthy – experts in scientific fields, including astronomy and mathematics. Although the State didn't avail itself often of the Society's advice, unless a clear benefit existed, there were links between the Royal Society and the Admiralty. The strongest examples of this were the establishment of the Royal School of Mathematics and the Royal Observatory, to assist in navigation, and Christ's Hospital, to train naval surgeons. As expeditions to new worlds returned with novel plants and other produce that, with scientific assistance, could be of service to the State, the Royal Society would gain greater influence and relevance. The Royal Society would be the chief promoter of the *Endeavour* voyage.

In many ways as a consequence of the Treaty of Paris, the British were now unintentionally grooming a perfect future explorer in Cook, and his crew, in *Dolphin*.

CHAPTER THREE

ENDEAVOUR, A GROUNDBREAKER
FOR THE BRITISH

The *Endeavour* voyage from 1768–71 was the big navigational breakthrough for the British. The lasting impact of the voyage was the prominence to which its leaders rose and the dynasty of maritime surveyors that resulted from their influence. It achieved for the British what the French would spend another fifty years building.

The Royal Society was the main proponent and financial underwriter of the *Endeavour* voyage. For the Society it was important that the transit of Venus in 1769 be correctly observed. This information could then be used to calculate the earth's distance from the sun. Observation of the transit was to be an international undertaking by a number of scientific institutions from Britain, France and other European countries, simultaneously, in a number of locations around the globe. The previous attempt in 1761 had been poorly performed, and the next transit after 1769 wouldn't occur until 1874. As national scientific ego was at stake the British were keen to ensure that their measurements, to be conducted somewhere in the Pacific, were the most accurate. Royal Society fellow Alexander Dalrymple was chiefly responsible for its planning. He knew nothing of Cook. He was also a senior official and hydrographer for the British East India Company and of some influence.

Dalrymple was convinced of the existence of *Terra Australis Incognita*. Like many of his contemporaries, he was a keen supporter of the theory that the existence of a large landmass in the South Pacific, south of Easter Island – and nowhere near Australia – was a necessary element of balance to ensure that the earth spun truly on its axis.

Dalrymple was also aware of information that most around him didn't know. Having acted as a quasi diplomat at the brief British occupation of Spanish Manila in 1762, prior to the Treaty of Paris, he appears to have had access to reports of Torres's previously secret glimpses of land south of New Guinea in 1606, separated by a narrow strait. The Spanish had jealously guarded this knowledge ever since, afraid of what it would mean for their settlements in the East Indies if enemy privateers got hold of it. The damage done to their colonies and bullion galleons over the centuries by the likes of Drake, Raleigh and, later, Anson were proof of this. It isn't known what else Dalrymple may have seen in Manila, and how much else the Spanish knew of Australia or *Terra Australis Incognita*. The Portuguese, rivals of Spain and keen explorers, were rumoured to have visited the Australian east coast long before. Certainly maps had been published asserting the existence of *Terra Australis Incognita*, as well as Torres Strait and what purported to be the east Australian landmass, but their reliability was questionable. Any records held by the Portuguese had been destroyed in the Lisbon earthquake of 1755.

Whatever the basis of his belief, Dalrymple was convinced that *Terra Australis Incognita* was there, somewhere south of Tahiti. Wallis had reported evidence of this, and Carteret had yet to return with his more pessimistic report – he was already believed lost for good. Dalrymple was also conscious of the Admiralty's interest in claiming this land before the French; the then current voyage of Bougainville had certainly focused French attention in that direction.

For a voyage to the Pacific, Dalrymple needed a good ship, and a good ship's master to sail it there, but he wanted to command the expedition – and bask in the glory of any new discoveries – himself. The Admiralty made it clear that if a naval ship was used, it was to be commanded by a naval officer. Dalrymple was no sailor. He was advised that although he couldn't direct command of the ship, he was welcome to go along as the scientific observer. Obstinate and proud, Dalrymple insisted that he either go as commander or not go at all. The Admiralty called his bluff. He didn't go at all. The astronomer Charles Green, a Royal Society appointee, replaced him as observer on the proposed expedition. The commander chosen was James Cook.

This was a surprising appointment. Following the Treaty of Paris, as during any time of peace, there was a surplus of unemployed and well-connected officers ashore on half-pay. Why was the relative newcomer, Cook, selected as commander of such a vessel – in place of not just any master, but the influential Dalrymple? This was probably less a matter of luck than of timing and ability. By then Cook was well known to the Admiralty as a competent sailor, navigator and surveyor, and carried the support of the Newfoundland commander-in-chief, Commodore Sir Hugh Palliser. The Admiralty was extremely influential with the Royal Society. The Royal Society also knew Cook independently, from his paper on the solar eclipse. But the Royal Society had the final word, in the form of an entourage of five additional 'savants' and their servants foisted on Cook at the last minute, as observers only. Of the five, only two would survive the voyage. At their head was young Joseph Banks, keen amateur botanist and a close friend of the Earl of Sandwich, the First Lord of the Admiralty. Banks was also a significant financier of the voyage, contributing even more towards it than King George III.

The salient features of *Endeavour* were that she had an experienced commander plus officers with the right kind of experience. From Wallis's *Dolphin* Cook obtained Lieutenant John Gore, a veteran of both the Wallis and Byron voyages, Master Robert Molyneaux, Master's Mate Dick Pickersgill, warrant officer and able seaman, respectively, Francis Wilkinson and Francis Haite. The remaining ex-*Dolphin* crew member, a midshipman from Byron's voyage, was Charles Clerke, a fascinating character who was to accompany Cook on all three voyages. His dry wit inspired as much loyalty in his men as his seamanship. For example, he punished native thieves in the Pacific not by flogging, cutting off earlobes or even shaving the whole head as others did, but instead by shaving one half of the head only. This produced an equally effective deterrent through ridicule rather than pain. During the Seven Years' War, as a young midshipman in action offshore, he had survived a fall from the topmast when it was shot away from his ship and he had plummeted to the sea. Despite his experience as a veteran of all British Pacific expeditions except Wallis's, he was late to join Cook's third voyage as captain of *Discovery* due to detention in an English debtors' prison, where he also contracted tuberculosis. He would take charge of the expedition after Cook's death but sadly succumb to the tuberculosis not long after, in Kamchatka.

Cook also had with him five men from his previous ship, the *Grenville*. They were used to the rigours of charting the Newfoundland coastline with him. So Cook could count on most of his senior officers, Gore, Clerke and Molyneaux, as experienced circumnavigators, and on a core section of his crew as capable, reliable and loyal. Even the lowly midshipmen were experienced. For example, it was young midshipman Jonathan Monkhouse's merchant sailing experience that was to save *Endeavour* and its crew when she was sinking after twenty-three hours aground on coral in the Great Barrier Reef and all other efforts to keep her afloat were failing. Monkhouse came up with the novel idea of fothering the ship with oakum and tar. Without men like these the voyage wouldn't have been the success it was.

As a result, Cook and the official astronomer, Green, recorded the transit of Venus, the prime object of the expedition – albeit inaccurately, due to the haze around the planet, apparently caused by the effect of the sun's heat on the tele-scopes used. In addition the expedition charted both islands of New Zealand, not visited by Europeans since Tasman, almost 130 years before, and delineated – rather than 'discovered' in the true sense, although that word will also be used from now on in this context – the east coast of Australia and its treacherous coral reefs. Extensive botanising produced a huge number of previously unseen and unknown plant and animal specimens. All in all it was a remarkable result, par-ticularly when compared to the relatively bare results generated by Byron and Wallis. It also provided the British with a respectable response to the published journals of Bougainville.

The voyage didn't give Dalrymple what he had wanted, as Cook hardly even touched on the area where Dalrymple thought *Terra Australis Incognita* should be. Cook also proved that wherever it might lie, New Zealand wasn't part of it. As regards

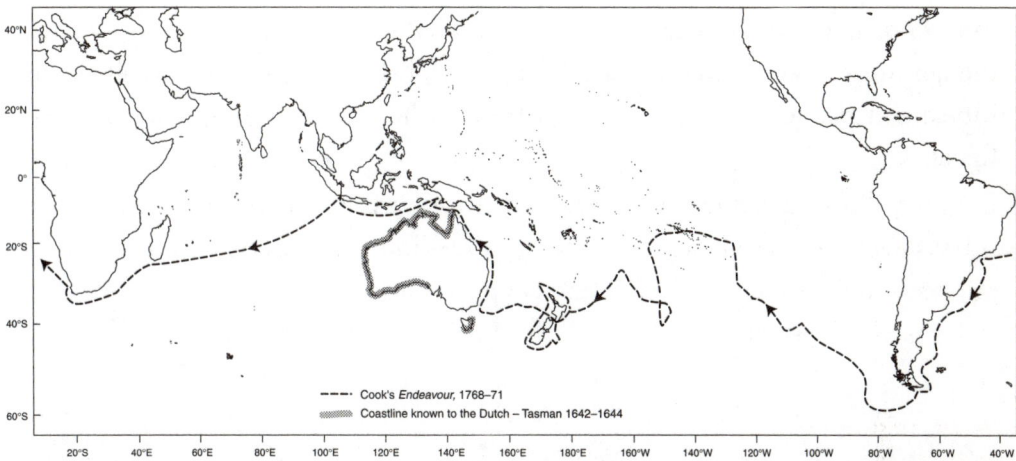

Cook's *Endeavour* track

the discovery of New South Wales, Dalrymple already knew the whereabouts of the landmass of New Holland, and wouldn't have been excited to know the exact delineation of its east coast.

Cook himself was less than effusive about his achievements, particularly on the Australian east coast. He wrote to his former employer – something of a father figure – John Walker of Whitby, saying with less modesty than one may have thought appropriate, 'I however have made no great discoveries yet I have exploar'd more of the Great South Sea than all that have gone before me so much that little remains now to be done to have a thorough knowledge of that part of the Globe.'[1]

Cook never returned to New South Wales. The *Endeavour* was sent off to the Falkland Islands and into obscurity as a storeship, to be ultimately sold into French service and rot away in the mud at Newport Rhode Island. The British were focused on developing their American and Indian possessions following the successes of the Seven Years' War. They had no interest in Australia.

Upon the safe return of *Endeavour*, the relevance of the voyage wasn't immediately grasped; it had achieved its scientific object, of interest to a few, and little more of obvious importance. Its importance would become apparent shortly after, though, for one individual in particular – Joseph Banks – and as a result of the subsequent rise to power he leveraged from it, he was to become a key historic figure.

CHAPTER FOUR

$\mathcal{A}$N 'ENLIGHTENED' AND ASCENDANT BANKS

Cook's career may have gone no further but for Joseph Banks, the rich young landowner and self-taught botanist from the voyage who had appeared on *Endeavour* at the last minute. Banks, not one to shy away in early life from publicising his accomplishments, rocketed to national prominence upon *Endeavour*'s return through his energetic promotion of the voyage's achievements – by suggesting them largely as his own – including the discovery of the tantalising delights of Tahiti.

Banks used his connections, both scientific and social, and through his friend – and First Lord of the Admiralty – the Earl of Sandwich, was able to ensure that the Admiralty wasn't offended. Journalist John Hawkesworth was commissioned at vast expense to write the narrative of the voyage from Cook's and Banks's journals while Cook was away on his second voyage. The press sensationalised much of what happened on the voyage, and Banks's involvement in it.

The private fascination of George III for agricultural and botanical pursuits, coupled with estimates of at least 1400 entirely new plant specimens from a total of over 30 000 gathered,[1] with which Banks and fellow botanist Solander had returned, ensured for Banks an introduction to the monarch.

From there, Banks was able to cultivate his value to the king in particular and the government in general. He was shortly afterwards appointed director of 'farmer George's' gardens at Kew, but this was only the first step.

Science was now being promoted as integral to the country's commercial advancement, and new scientific discoveries were seen as a key. The age of 'Enlightenment', as it would become known, was sweeping Britain and Europe, and – thanks to *Endeavour* – Banks would become one of its most influential proponents.

Banks used his profile from the *Endeavour* voyage to become one of the high priests of an applied form of scientific Enlightenment, at a time when science as a separate profession didn't really exist outside specialist fields such as astronomy, which was provided with government support. Scientists of that age in Britain were usually so only as a by-product of a more lucrative profession, such as medicine, or were independently wealthy and followed the pursuit of science more as a hobby, as 'dilettanti', the peak of which would be a fellowship in the Royal Society. The dilettanti slowly coalesced to a body with more scientific discipline and rigour as Enlightenment set in.

'Enlightenment' has been used as a term to describe the closing decades of the eighteenth century in a number of very different ways, from changes in political freedoms to liberalisation of scientific thought. For our purposes, Enlightenment involved the development of rational principles derived from fact or experiment, rather than religion, to explain observed phenomena. This was a significant change, in an era where church theology was often the starting point for such discussion, where man was definitely a creation of God, and the theories of Charles Darwin suggesting otherwise were still one hundred years away. At its most useful, in Banks's mind, Enlightenment represented the harnessing of scientific knowledge in furthering the interests of State, particularly trade.[2]

The raw materials for many of Britain's manufacturing industries would grow only in tropical climates, more often than not in areas not controlled or colonised by Britain. Examples were flax for its ships, dyes, fine cottons and wools for the textile industry, and sugar and spices for food and beverages. Supply was always subject to the vagaries of war and political and economic intrigue. If seeds could be transported from their foreign origin to British-controlled environments, such as Jamaica, and developed for commercial use in British colonies, this risk could be avoided.

Transshipment of plants between countries and their propagation in botanical gardens was one of the better-known examples of Enlightenment in action. Another was smuggling Spanish fine-wool merino sheep from Spain to Britain to raise a flock of healthy breeding stock for sale to British breeders – and to others, such as John Macarthur. Banks played a pivotal role in both initiatives.

This form of applied science wasn't totally new. The Society of Arts, Manufactures and Commerce had been established in 1754 on a similar theme, and offered bounties and other rewards to those who could, for example, transplant cinnamon from its native source to successfully grow in the West Indies. Banks brought new energy to this ideal, particularly with the large number of natural history discoveries he'd made during the *Endeavour* voyage. He was elected as president of the influential Royal Society in 1778, and would hold the position until his death in 1820. It gave him immense power and a platform to further these ideas.

Banks realised early on the changes that were taking place in the developing industrial world of Britain and the need to preserve and foster its manufacturing sector. In this he was way ahead of his time, as the full effect of the industrial revolution and its steam engines, looms and machinery wouldn't become apparent to many until the next century. For example, disciples of applied Enlightenment like Banks believed that colonial outposts such as India should supply unprocessed raw materials, dyes and medicines to Britain and support Britain as processor and manufacturer. They certainly shouldn't compete as manufacturer themselves. Plant transshipment through botanic gardens would only produce a wider selection of raw materials from which to manufacture a wider product range in Britain. If these outposts could also be encouraged to trade raw materials with other markets, such as China, then an even more serious problem for Britain – the outflow of British bullion needed to support these outposts or to buy trade goods from China – could be reduced. The key was encouraging the development of raw material sources in colonial outposts such as these and ensuring their supply to Britain. In this way Britain, despite its adverse climate, could access the benefits of produce from, potentially, anywhere in the world.

More recent attempts at seed transplantation by the Society of the Arts, Manufactures and Commerce had been thwarted by the American War of Independence, particularly once the French Navy had sided with the colonists. Indeed it seems that the French were more advanced in this field than the British. They'd already transplanted nutmegs, cloves and East Indian breadfruit from the East Indies to Mauritius and the French West Indies. Further, French vessels had been captured during the American War of Independence in the Caribbean carrying breadfruit from Mauritius. It was probably almost inevitable that Banks became such an ardent supporter of the transplanting process.

Banks summarised a more benevolent aspect of his plans to the secretary of war, George Young, in 1787 in relation to breadfruit:

> … to exchange between the East and the West Indies the productions of nature useful for the support of mankind, that are at present confined to one or the other of them, to increase by adding this variety, the real Quantity of the produce of both Countrys & by that means their population …[3]

Botanical gardens were established at Calcutta, Madras, Bombay and St Helena between 1786 and 1791 with that aim – cultivating useful plants from other environments. St Helena in particular was a useful staging point between transfers, where plants could be restored from the adverse effects of salt, wind and cold during seaborne shipment or await favourable seasons for travel. Prompted by Banks, the East India Company promoted the development of these gardens by the provision of funds and land. Banks was usually responsible for finding appropriate botanists to run them. The most famous example of this program was the transshipment of breadfruit in the *Bounty* from Tahiti to the West Indies.

Over his lifetime Banks developed a huge network of natural history collectors throughout the globe who would send him seeds, dried plant specimens and other items of animal or mineral interest. Through his network Banks advised the government on, and in some cases more actively promoted, similar schemes, including the introduction of Chinese tea for cultivation in Bengal, the growing of flax in India – the basis of the future jute industry – and, as mentioned, the smuggling of merino sheep from Spain for a flock at Kew Gardens in order to improve British wool breeds.

Many of these schemes were only made possible because of the voyages of the early navigators, such as Cook. They made the Western world aware of its neighbouring undiscovered lands and their produce. To Banks, the navigators were an important part of the Enlightenment movement. Without them – especially without their charts – without their well-commanded ships and inquisitive commanders, and without the Banks-appointed botanists, artists, geologists and naturalists aboard to all help search out new and unknown material, Enlightenment couldn't achieve all its goals.

Also, in a sense, as wars intensified towards the end of the eighteenth century, the navigators couldn't achieve their objectives safely without the help of the scientific minds of the Enlightenment. By this time, pursuit of scientific knowledge had generally become a universally supported goal among European powers, who issued passports of free passage to enemy ships provided that they were engaged on scientific expeditions, rather than connected with the conduct of war, trade, contraband or the carrying of enemy government dispatches.

Banks was the first to arrange such a passport. During the American War of Independence he used his influence through his friend Benjamin Franklin to procure a passport from the American rebels for Cook's third voyage. The French, as American allies, issued a similar passport, instructing their ships not only to give Cook free passage but also supplies and assistance if required. In the war-torn closing decades of the eighteenth century, this arrangement gave additional incentive to voyages of a scientific nature. If a nation was interested in launching an expedition to find new trade routes or possible bases for trade or settlement, as had been the case with the earlier voyages of Byron or Bougainville, it could only do so with the immunity of a passport if it had this scientific character. This again meant that the involvement of an informed, apolitical person with a large network in the various sciences, such as Banks, was even more essential to the navigators' voyages.

Banks thus became the unofficial, apolitical, massively powerful and wholly indispensable patron to the British navigators.

Endeavour was the starting point for Banks. Through his efforts, and the further involvement of Cook, a dynasty of British navigators was forming.

CHAPTER FIVE

COOK'S SCHOOL

The public interest created by Banks in the *Endeavour* voyage, as well as Cook's own charts, ensured for Cook a second voyage. This time its aim was to establish for certain the existence – or not – of *Terra Australis Incognita*, as Dalrymple so desperately desired.

Cook's second voyage

Two ships, *Resolution* and *Adventure*, were to be used. However, the slow demise of Dalrymple's power was evidenced by the fact that the promoter of this trip was instead Joseph Banks. But this time the young Banks pushed things too far with the Admiralty, and insisted on being accompanied on the voyage by fourteen scientists, together with the painter Zoffany, two French horn players and four footmen. In order to fit everyone in, Cook was required to give up the great cabin entirely – he'd shared this with Banks on *Endeavour* – and was to live in a special roundhouse constructed on *Resolution*'s deck. The result of all these changes was that *Resolution* was top-heavy and unsailable.

The Admiralty immediately countermanded Banks, tearing out the roundhouse and additional accommodation that Banks had insisted on for his large contingent. This infuriated Banks to such a degree that despite his newly won influence, and with a dockside tantrum in front of the *Resolution*'s crew to emphasise the point, he withdrew all support for the expedition, retiring instead to Iceland with his retinue, a different ship and one of Cook's best officers, John Gore, in command. It was almost a repetition of the Dalrymple affair before the *Endeavour* voyage, and vindicated the Admiralty's view on keeping civilians well away from command of navy ships on these expeditions.

Cook persevered with his orders, and executed his second voyage from 1772–1775, showing conclusively that there was no *Terra Australis Incognita*. Banks was replaced by Johann Reinhold Forster and his son, George. Despite the dearth of new discoveries, this expedition should rank at least equally with Cook's more famous first voyage in terms of achievement and seamanship. Cook penetrated far deeper into the Antarctic Circle than those before him, to 71º10′, and circumnavigated the Antarctic continent. When the decision to retire from the high latitudes was finally made, young seaman George Vancouver climbed out on *Resolution*'s bowsprit as she wore away from the South Pole, waving his hat and yelling, 'ne plus ultra'. He boasted thereafter (in fact, for the rest of his short life) that he'd been closer to the South Pole than any other man on earth – by the length of the bowsprit.

No-one would get closer than young Vancouver for another fifty years or so, when James Weddell penetrated only slightly further south in 1823. Unfortunately, Cook returned to a Britain facing revolution in the American colonies and wars with both France and Spain, and was unable to deliver discoveries of political or economic advantage. There was no interest in the discovery of a huge frozen continent.

Cook's third voyage

For his third voyage in 1776–80, Cook was saddled with a chestnut that neither he nor any other sailor would crack: he was ordered to seek a north-west passage across the top of North America. Had this been possible, it would have provided great advantage to Britain, bypassing the troublesome waters of French, Spanish,

Dutch and – shortly – American possession. It was a dangerous assignment; the tenacious Cook and his ships may well have not returned, lost instead amid the polar ice. This fate awaited a later student of Flinders, John Franklin. *Resolution* and *Discovery* were probably saved from this horror by Cook's death in Hawaii on St Valentine's Day, 1779, after a bungled hostage-taking episode over a stolen boat.

The voyage produced the newly discovered Hawaiian Islands, as well as the realisation of the huge exploitable potential of the North American fur trade in Asian markets.

The 'school' established

Cook left a powerful legacy, and not only in the charts he produced. Perhaps of more importance was his contribution to the minds and wills of the officers and seamen who learnt from him in constructing those charts over almost eight years at sea, as well as those who would in the future learn from them – 'Cook's school'.

Take the devoted and fiercely loyal young master of the *Resolution* on Cook's third voyage, William Bligh. His temperament wasn't unlike Cook's; he was an equally thoroughgoing seaman, and even his background was similar. He started as a volunteer seaman with no connections and worked his way up through the ranks, teaching himself further skills of charting and astronomy as he went. Bligh had sailed Cook's ship, helped chart his coasts and learnt much from him. He would survive three mutinies and two major sea battles, and conduct voyages of exploration of his own over his naval career.

Bligh would go on to influence and educate Midshipman Matthew Flinders on the second breadfruit voyage on *Providence* in 1792. Flinders in turn would take his cousin, John Franklin, as a young midshipman on the *Investigator* in 1801. Their influence played a part in the career of Phillip Parker King, who charted the northern coasts of Australia in 1818 and during the 1820s. Rear-Admiral King would support and promote the efforts of later surveyors of the nineteenth century, in particular FitzRoy, Wickham and Stokes – each commanders of Darwin's *Beagle* – and after them Stanley and Denham. Cook's school set up a solid tradition of British navigators.

From Cook's fellow officers, supernumeraries and crew, two names emerge that that stand out. These are Bligh and Banks. Others, such as Riou, Broughton, Colnett, Puget, Dixon and Portlock, would make their marks in history in ways that don't touch on this story. There is, however, one other relevant name, George Vancouver, the lively young able seaman on the *Resolution* on Cook's second voyage. We last saw him near the South Pole swinging from the jib boom of *Resolution*'s bowsprit yelling 'ne plus ultra' as she turned north.

Vancouver was a midshipman in *Discovery* on Cook's third voyage. As such he would've had full benefit of the navigational teaching of Cook as well as of Charles Clerke, captain of *Discovery*. Vancouver's contribution would come with his own expedition, of marathon length, from 1791 to 1795. He was sent, in command of His Majesty's ships *Discovery* – a newer, improved version of Cook's support ship – and the smaller *Chatham*, to receive formal surrender from the Spanish of the recently seized English settlement at Nootka Sound on the American west coast. He was also ordered to survey as much of the American coastline as possible and to further explore the south and west. On the voyage out in 1791, he charted a portion of Western Australia, discovering and charting George III's Sound (Albany). He then headed on to New Zealand and to North America.

Vancouver's surveys were to be some of the most scientific and accurately done over a huge coastline at that time, with Vancouver himself carrying out many of the astronomical observations. He proved that there was no major seaway from that coast into inland North America, and surveyed the Hawaiian Islands. At Nootka Sound, he also dealt with some highly sensitive political misunderstandings on the part of his government with the Spanish. And he persuaded the Hawaiians to cede to British protection; unfortunately for the British, this was never ratified – the British Government of the day was more concerned with the after-effects of the French Revolution. On top of this, over a 57-month voyage, *Discovery* lost only one man from disease, and *Chatham* none.

Despite having so many brilliant qualities, Vancouver seems to have shared all the elements of the tyrant that history has since landed on Bligh. Although this may have had something to do with a thyroid deficiency from which it appears he suffered, one wonders whether this was also a result of, rather than in spite of, Cook's example on that third voyage, as we shall see. During Vancouver's long voyage, it appears that nearly half of his crew were flogged, and Archibald Menzies, the botanist appointed by Banks for the trip, was placed under arrest and confined to his cabin for a period in the closing stages of the voyage. But his treatment of one midshipman in particular ensured the complete destruction of Vancouver's career. The offending individual was left behind by Vancouver at Hawaii to find a ship back to Britain. He later accused Vancouver of having him flogged during the voyage – not an unusual punishment for an ordinary seaman but never for anyone entitled to pace the quarterdeck, even if only a midshipman. This alone would probably not have hurt Vancouver. It seems as though the midshipman was hardly leadership material; he later left the navy and was killed in the last of several duels he fought. What did damage Vancouver was that the sailor in question was Thomas Pitt, cousin of the then prime minister, brother-in-law to the foreign secretary, and cousin to the First Lord of the Admiralty, on whose favours Vancouver depended for promotion, and the future Baron Camelford. The resultant criticisms through the well-connected Pitt family, Banks – Archibald Menzies was one of his favourites – and others ensured that, despite promotion to post-captain, Vancouver's career went nowhere after his voyage, notwithstanding his brilliance as a surveyor and sailor. Already a sick man during his lengthy expedition, he died in 1798.

Cook's voyages produced a lineage of young, experienced sailors well trained in basic hydrography. This lineage also established the foundations for their powerful patron, Joseph Banks. All this started with *Endeavour*.

CHAPTER SIX

DISASTER FOR BRITAIN; OPPORTUNITY FOR BANKS

While France had been recovering from the effects of the Treaty of Paris of 1763 and attempting to re-establish its international trading base, the British had embedded their influence globally. In India, Robert Clive had consolidated British gains in Bengal; British sovereignty had been established over the eastern seaboard of America; and a healthy trade, chiefly in sugar, flowed from its expanded possessions in the West Indies. Britain had little need of new colonies. At the time of Cook's second voyage in 1772, Britain's position was, from an international point of view, comfortable and relaxed.

The loss of the colonies

This rapidly altered in 1775, on a date which history marks as 19 April, when the first shots were exchanged at Lexington between British troops and local militia, sparking the American War of Independence (1775–83). The first years of the war saw sufficient British successes to perhaps provoke complacency at home. Cook's third voyage was dispatched during this period, with instructions to chart the north-western American coastline and find the North-West Passage, both of which would still be useful to the British-controlled colonies.

This attitude must have changed considerably following the British losses at the Battle of Saratoga in autumn of 1777. These disasters persuaded the French to ally with America in 1778 as a means of regaining some possessions removed by the Treaty of Paris. Alliances followed quickly thereafter between America and the Dutch and the Spanish. Britain was suddenly friendless in continental Europe. Its naval power was nevertheless required to maintain and supply its beleaguered North American armies on the one hand and, on the other, to keep open the trade and supply routes with which it had been so comfortable and relaxed in the recent past. This it couldn't do. A combined land and sea assault on Cornwallis's besieged British troops at Yorktown by Washington's army and the French Navy forced the British surrender in 1781, and sealed Britain's defeat in the American War of Independence.

The loss of the British American colonies and the treaties of 1783 that followed would deal shocks to Britain similar to what France had felt almost ten years before, following the Treaty of Paris. The political climate of Britain was thrown into turmoil; the government of the day was expelled, taking with it Banks's great patron, John Montague, Earl of Sandwich and First Lord of the Admiralty. The new government of the Earl of Shelburne and his understudy, William Pitt the younger, instigated major economic and bureaucratic reforms. Pitt believed that significant change was necessary if disasters such as the loss of the American colonies were to be avoided in the future. Pitt's focus on accurate accounting and accountability, reliance on external objective and professional advice and efficient administration without sinecures and stupidities would pay off when the stress of the Revolutionary and Napoleonic wars hit Britain shortly afterwards.

Most relevant of these reforms for the navigators was the abolition of the Board of Trade and Plantations and its replacement in 1784 with the Privy Council Committee for Trade, run by Charles Jenkinson, who became in 1786 Lord Hawkesbury. The Board of Trade had become a large, unwieldy body with prime responsibility for the American colonies. It was to take considerable blame for the failure of the British Government to recognise the depth or seriousness of the

American rebellion or the extent to which it would engulf that territory. The immediate impact of its abolition in 1782 was that there ceased to be any advisory arm of government with a distinct interest in science, exploration or the colonies.[1] This created a major opportunity for the recently elected president of the Royal Society and confidant of King George III following the *Endeavour* voyage, the 'Enlightened' Joseph Banks.

A step for Banks

Banks was well acquainted with Lord Hawkesbury. The latter's newly created and powerful Committee for Trade, as well as other arms of government, was in continual need of scientific advice. With his position, influence and international reputation, Banks was in a unique position of de facto power on matters scientific, exploratory and colonial, which ensured for him an ongoing role in British foreign policy. As a man of science he was apolitical, both locally and internationally. This meant that he was regarded by all as something of a neutral. He could converse with both sides of parliament in Britain. In France, republican and royalist alike regarded him with respect as president of the Royal Society, rather than with hatred as an enemy of France, even through the darkest days of the Revolution. Napoléon Bonaparte was an enthusiastic supporter of Banks – the sentiments weren't reciprocated – and had invited him to Paris despite the hostilities between their nations. Banks was also a great correspondent of American republican Benjamin Franklin, through whom he initiated the practice of passport issuance by arranging a passport for Cook's third voyage during the American War of Independence. This set a precedent for the voyages of La Pérouse, d'Entrecasteaux, Baudin and Flinders.

Understandably, Banks was only a de facto adviser to the British Government. Anything more formal would have jeopardised his influence and power base with a change of government. Official responsibility for the colonies still rested with the Home Office, which had its hands full for the remainder of the eighteenth century managing fallout from loss of the American colonies and the French Republican and Napoleonic wars.

Banks's political neutrality was such that he was trusted by whoever was in power at the time until the latter part of the first decade of the nineteenth century.

Both Hawkesbury and Evan Nepean, First Secretary for the Home Office, would come to rely heavily on Banks for advice, to the extent that we see Banks drafting instructions for Flinders and the *Investigator* in 1800 and appointing and instructing a series of governors to New South Wales until 1806. We read Evan Nepean's often quoted words to Banks, '… any proposal you make will be approved. The whole is left entirely to your decision.'[2] It's indicative of Banks's political and diplomatic skills that he enjoyed a similar position with the East India Company.

Banks was the linchpin for the navigators. Here was a patron, provided it was the right sort of voyage, with significant influence with the government – regardless of its persuasion – the monarch, the Royal Society, foreign enemies and anyone else who mattered. He offered constant support to the navigators in order to advance his own versions of 'Enlightenment'. The French had nothing like this form of apolitical support for their navigators. Bougainville and Fleurieu had some influence, of course, but the trauma caused by the French Revolution and the uncertainties that followed as Napoléon slowly climbed to power denied them the relatively stable framework within which Banks was able to work for the benefit of the British navigators. Banks would only start to lose this unique position with the demise of influential contacts following the death of Hawkesbury in 1808 and the onset of George III's madness from porphyria in 1810.

A first fleet

In 1787 a fleet of transport ships and escorts sailed from Portsmouth. From the navigators' perspective this was the most resounding consequence of the American war. It's now known as the First Fleet, and was sent to form a penal colony in New South Wales. Given the distinct British distaste for creating further colonies without clear benefit following the loss of the American colonies, what had lead to this result?

The American colonies had provided a useful receptacle for transportation of convicts from Britain, but this was now lost following the American war. Britain was left with overcrowded jails and prison hulks. Mass migrations from country to city in search of work following the onset of the industrial revolution and food shortages from combinations of crop failure, wartime naval blockade and the demands of British armies in foreign fields added to these stresses. Demobilised soldiers returning home from the American war only added to the ranks of Britain's unemployed, impoverished and potentially criminal class.

As early as 1779, Banks had proposed New South Wales as a suitable convict colony. This proposal went nowhere. New South Wales was too far away to be efficiently supplied from England, and alternative sites were available. These alternatives weren't pursued, and the issue arose again in mid-1783 when ex-*Endeavour* midshipman James Matra, now unemployed, had written to Banks asking for information on any proposed settlement. Within a month, Matra – in consultation with Banks – had proposed a scheme to the government for settlement of New South Wales with displaced royalists from America. Again the proposal was shelved and the royalists went elsewhere.

By 1785 matters had again come to a head, and the Home Office was seriously considering an alternative site in Africa for a penal settlement. But after a fruitless survey of the proposed site in early 1786, a decision was made only a matter of months later in August 1786 that favoured an alternative location, Botany Bay. Botany Bay was clearly not preferred, given its distance from Britain, difficulties with communication and the likely costs of maintaining the colony, but by then matters had reached a point of desperation, with no other sites available.

It's also possible, as many historians have since argued, that Botany Bay was chosen as part of a much wider British political agenda, in order to establish a British Pacific base for trade. This would allow the British to access the northwestern American fur trade, as well as Asian markets. It would also provide a much-needed new base for the British whaling industry, currently facing increased competition and declining stocks in the Atlantic. Then again, perhaps the decision encompassed a convenient solution to a number of the pressing trade worries, as well as the major issue of penal settlement. In any case, these additional concerns may well still have required the involvement of Banks. The decision to proceed with the settlement, the most profound of any affecting the modern history of Australia, can therefore trace its roots to original participants of Cook's voyages, especially Banks.

Ultimately the loss of Britain's American colonies was good for its navigators and for Banks. It led to Banks consolidating his power base in the British

Government, to the settlement of Port Jackson, and to a demand for coastal and inland exploration and surveys in Australia.

Indirectly the American war wasn't beneficial for France. In allying itself with the rebel government during that war, at a relatively precarious time in her own social history, France's economic woes had worsened considerably. By 1787 the French controller of finances proposed a system of taxation to reduce France's debt. This, together with the popular concept of democratic rights alongside which France had fought in the American wars, provided sparks that would lead to the formation of the National Constituent Assembly, and ultimately the French Revolution.

The French Revolution produced fresh and intense conflict between the British, the French republicans and other European powers following King Louis XVI's execution and the commencement of the Revolutionary Wars in 1793. This conflict brought about untenable turmoil among the French navigators. It tore apart one major French expedition, that of Bruny d'Entrecasteaux. Political unrest arising from the Napoleonic Wars ensured that the succeeding expedition, of Nicolas Baudin, also failed to receive the credit it merited. Any likelihood of the French creating or maintaining a line of experienced maritime explorers to rival 'Cook's school' before the turn of the century was lost.

CHAPTER SEVEN

Essentials for the navigators' expeditions

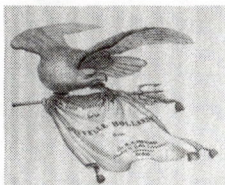

The Treaty of Paris, the American War of Independence, the French Revolution and the wars that followed set the high-level political background for the various expeditions from France and Britain that would visit the Pacific over the last thirty years or so of the eighteenth century.

This general political backdrop was only one factor in the success or failure of these voyages. Success was ultimately down to the individuals in command, the navigators themselves, and their resources – the ships, the crew, the anchors and the ship's boats. It's at this level that the outcome of the entire venture could hang on a stroke of luck, a gamble on the wind shifts or the management of the crew and supernumeraries.

What ingredients had made the *Endeavour* voyage such a success? Why were Cook's crews so scurvy-free, when he actually had no idea what, in particular, kept scurvy at bay? Why did some of the French voyages have scurvy in their ships while adhering closely to his remedies? Why did Cook resort to almost barbaric means to exert his authority on his last voyage? Why did so few of the French navigators survive their expeditions, when the British navigators did? There are no clear answers, except to consider the large number of variables that could mean the difference between failure and success.

The ship

Selecting the right vessel, particularly best design and condition, was of critical importance. Flinders' expedition on *Investigator* had to be prematurely terminated on account of the ship's decay, and the unseaworthiness of his replacement vessel ultimately led to his own demise. Conversely, Cook's voyages were highly success-ful due to an unexpected decision by the Admiralty early on, involving what entrepreneurs today would term 'significant technology risk' in the choice of ship.

Cook's *Endeavour* was a Whitby collier, colloquially known as a 'cat', or 'cat built'. Her broad bluff bows (for strength), wide beam (for stowage room), rela-tively shallow draught and flat bottom (to negotiate shallows and tolerate ground-ing) were novel for this type of voyage. These features allowed far more flexibility and durability for voyages of discovery than the likes of the faster *Dolphin* or the tiny *Swallow* that had been employed to date by the Admiralty. *Endeavour*'s light rigging, lacking the aft top-gallant mast and yard usually seen on square riggers of the day, was less complicated and needed fewer crew to manage, as speed wasn't one of *Endeavour*'s strong suits. Given the hammering she received on the Australian Great Barrier Reef, particularly when aground for twenty-three hours off Cape Tribulation, Cook probably wouldn't have survived in any other sort of vessel. It's no wonder that the ships for all Cook's voyages, *Resolution*, *Adventure* and *Discovery*, though bigger than *Endeavour*, were Whitby cats.

Selection of ship was the first component. Then there was the constant task of maintaining her. Seagoing craft were essentially organic things, composed (as Ray Parkin pens it in relation to *Endeavour*[1]) entirely of trees and grass, with a mini-mum of iron, which were always in need of upkeep. Her rigging, all of hemp rope, would stretch and chafe and required constant attention. Old rope was never thrown out. It would be unwound and respun as spun yarn for less stress-ful tasks or otherwise left as 'junk'. In its final unpicked form as oakum, com-bined with tar, it would be used to caulk and seal the hull. A large supply of old rope was useful to repair split yards and masts until replacements could be found.

In fact the poor state of *Endeavour*'s rigging and sails after a year at sea was indirectly responsible for the discovery of the east coast of Australia. There'd been nothing in Cook's orders directing him to Australia. After completing the circumnavigation of New Zealand, he had to decide which route to take for the return voyage to England – either west via unknown shores, Batavia and the Cape of Good Hope but with favourable winds, or east via the Horn and its terrible, freezing storms. He personally favoured the eastern route across the Pacific to the Horn, away from Australia – a course which would have given him the opportunity to settle whether or not Dalrymple's *Terra Australis Incognita* existed. But when Cook put the choice to a council of his officers, it was the state of the ship, particularly her worn rigging and sails, that ultimately decided him against this course. Cook accepted that these wouldn't stand up to the freezing storms and high seas in winter off Cape Horn and decided to follow his experienced officers' advice to sail west, rather than his own ambitions. Again, the counsel of the more experienced circumnavigators from *Dolphin* was invaluable.

The ship's boats

The ship's boats were essential. They were a lifeline for the crew, providing the only means of collecting wood and fresh water, constantly needed for the crew's sustenance on a long voyage. They were also relied upon to seek out safe anchorages and to land shore parties for surveying and botanising, particularly when hazards like shoals and unfavourable winds would allow access to nothing bigger or less manoeuvrable. There were no docks or jetties in the Pacific then.

The boats were vital for exploring bays, potential river mouths, sounding depths inshore for survey purposes, finding safe passages through coral and sand, and for laying anchors to allow the ship to be winched out of danger. Loss of a large boat was a major problem for an expedition. A justification for Baudin's early diversion to Timor from Western Australia, which resulted in much death on his ship, was to replace a lost cutter. And it was Cook's overreaction to the theft of a cutter by Hawaiians that led to his death.

The anchors

As ships of the day had minimal ability to sail close into the wind, anchors were essential security to avoid being blown ashore; they also provided a means of winching a vessel away from other dangers, especially along Australia's shallow northern coastline.

One of the biggest difficulties was ensuring that an anchor was dropped onto a sea bottom, preferably clean sand, that wouldn't allow it to drag and from which it could be raised without snagging or fouling.

Anchors were often poorly made, thanks to the scandalous behaviour of naval contractors of the day in all nations, and would frequently be drawn up after more than an hour's labour at the windlass with flukes bent or broken, making them next to useless. On a long voyage with little chance for resupply, spares could soon be exhausted.

Baudin constantly complained of his junior officers' negligence in losing or damaging anchors. On the west Australian coastline his consort vessel, *Naturaliste*, lost four anchors. Twenty years later Phillip Parker King was forced to terminate his survey along the north of that coastline because of anchor loss. The damage to and loss of anchors in the Great Barrier Reef were factors that forced Flinders to send his consort vessel, the *Lady Nelson*, back to Port Jackson. The anchor loss so concerned the crew of *Lady Nelson* that they made a replacement from a tree and two small cannon, which, legend has it, unfortunately floated when they needed it most.

Longitude

Accurate navigation was still in its infancy when *Endeavour* sailed. Although latitude could be easily ascertained with clear weather, a reliable means of determining longitude at sea had yet to be perfected. In 1714 *The Longitude Act* was passed, under which Parliament promised a prize of £20 000 to anyone who could invent a practicable means of measuring longitude at sea.

By the time of *Endeavour*'s departure for the Pacific in 1768, a solution was still unavailable. John Harrison had by then invented a sufficiently reliable chronometer for the purpose, but the British Board of Longitude wanted to develop a smaller, cheaper version.

Cook relied on other methods of navigation, using the astronomer royal's recently developed lunar tables and his own astronomical observations to determine his longitude. Given the flaws inherent in this approach, its dependence on clear weather for observations, and some error in the lunar tables themselves, it's impressive that he accurately charted as much as he did. Indeed Flinders spoke in almost shocked terms in his narrative thirty years later as he described how, with the benefit of a chronometer – in fact Flinders had three – he'd established that the charts of the 'immortal Cook' were inaccurate, sometimes by more than 100 kilometres.

Tasman's 140-year-old charts of parts of the Australian coastline couldn't be relied upon, but not for this reason alone. The lack of accompanying narrative, and the Dutch habit of deliberate concealment or else revealing misleading information in order to preserve the trading monopoly of the Dutch East Indies Company was legendary, leaving doubt as to who the charts' draughtsman actually was, as well as their accuracy.

It was also inaccurate recording of Pitcairn Island, both as recorded by its discoverer, Carteret, and his publisher, that gave Fletcher Christian and the *Bounty* mutineers their salvation. Carteret's longitude was out by about 3 degrees, and Hawkesworth, Carteret's publisher, had misrepresented his latitude by around 5 degrees, making Pitcairn's recorded position quite inaccurate and ensuring the mutineers almost twenty years of undisturbed occupation.

Hydrographic practice generally

Hydrography, the mapping of the earth's surface waters and measurement of its depths, was a relatively young skill when Cook and Samuel Holland had their discussions on the banks of the St Lawrence River. Generally, land-based surveying methods were adapted to the water. The most basic method of mapping a shoreline was with simple triangulation. This involved measuring a given length between two clear markers, for example between two poles along a beach, for use as a base line. Bearings would be taken from other points along the shore on each of the markers and from boats sounding inshore. Simple trigonometry then produced distances between the points at which bearings had been taken, and the base line and its markers. A chart could then be constructed combining this information. Soundings were carried out at timed and measured distances by dropping a leadline on string to the sea bottom and noting the length of string let out until the bottom was reached. The nature of the sea bottom was also recorded via the tallow or wax tip of the leadline to which sand or gravel would stick. This was also important information for determining an area's value as an anchorage. The difference between a fine sandy sea bottom and a rough rocky one could mean an insecure, damaged or lost anchor.

The more bearings taken, the more accurate the chart. All bearings would need to be adjusted for the earth's magnetic variation because the earth isn't a perfect sphere, meaning that true north differed from magnetic north. Also, as Flinders would discover, adjustment was required for the magnetic distortion caused by the effect of iron nearby, such as ships' cannon, if bearings were taken on board ship.

If time, weather, water depth, winds or the local inhabitants meant that baseline triangulation couldn't be used, and particularly on an unknown shoreline, navigators often resorted to running surveys, a far less accurate method, by which bearings were taken from offshore of major features, such as headlands, as a survey vessel proceeded along a coastline without stopping. These were compared with data from the ship's logline as to speed. Cook, Flinders and Vancouver did much of their work in this way. Running surveys would be later discouraged by the British Hydrographic Office as more detailed and accurate surveys of known shorelines became necessary.

The shape of the coastline, whether derived from a fixed or running survey, then had to be given global relevance by use of points of latitude and longitude. Latitude could be easily measured with sextant but, as noted above, longitude was far more problematic before the days of reliable chronometers.

Longitude, however measured, also varied between countries. The British measured longitude starting from a position of zero degrees at a meridian running through Greenwich in London. The French measured their base meridian at Paris, meaning that the French and British charts always had slightly different longitudinal positions for any point. The French had at one stage suggested a common meridian be used at Tenerife, but the question wasn't resolved until the early nineteenth century, at which stage the dominance of British sea power was reflected by the use of the Greenwich Meridian as the global longitudinal base.

A navigator therefore required a keen mathematical mind, a great deal of patience and sailing skill to be able to conduct running or land-based hydrographic surveys. Most naval officers had a degree of mathematical ability. They needed this to calculate latitude and longitude for basic navigation in any case, as dead reckoning – estimating a ship's position by educated guesswork, using the direction of travel and estimated boat speed – was never totally accurate. They acquired this knowledge as junior-ranking midshipmen (as Flinders did) or warrant officers (like Cook and Bligh) under the tuition of a more experienced commander or senior officer.

However, it was a big step to develop this knowledge for accurate hydrographic use, particularly when that science itself was in such early stages of development. This was why the exceptional mathematical talents of individuals like Cook, Bligh, Vancouver and particularly Flinders enabled them to achieve so much. During the eighteenth century the French commanders tended to rely almost entirely on civilian subordinates for hydrographic work. For example, Baudin, a skilled navigator and sailor, was no hydrographer, depending on others to do this work. His naval subordinate, Louis de Freycinet, would ultimately develop sufficient skills during Baudin's expedition to later conduct his own hydrographical expedition without civilians in the early nineteenth century, but almost fifty years after Cook.

Consorts

Often the navigators worked with more than one ship. Around the shallow and dangerous uncharted Australian coastline, they thought it safer if there were two ships rather than one – to assist each other in times of danger, to carry additional supplies and support, and also to share the burden of the survey work. If one of the vessels was more manoeuvrable and able to handle shoals and shifting winds, it was likely to be of much assistance on Australia's coastlines as it could carry out the more accurate inshore survey work that was too dangerous for the bigger vessel, or scout ahead for reefs and shoals which, while not a threat to its shallow draught, could wreck the bigger support ship. The cramped and overcrowded *Bounty* may not have had its mutiny, nor *Endeavour* its near-death experiences on the Barrier Reef off Cooktown, had a consort been present.

In theory the idea sounded good but in practice it was hard to execute successfully. Each vessel needed commanders who, although capable seamen and each totally able to command independently, could also happily operate together, with one subject to the orders and directions of the other. Cook's ambitions on his second voyage were occasionally frustrated by the more cautious approach of his consort commander, Tobias Furneaux; he found himself constantly missing rendezvous or otherwise mistaking his commander's intentions. It was Furneaux who perpetuated the theory that there was no Bass Strait after sailing north from Van Diemen's Land, although at some distance offshore and drawing this conclusion from winds and tides rather than his own observations. Cook supported him in his view but may have felt more than a little frustrated that he didn't actually go and find the answer.

Likewise the French had their difficulties. Baudin's consort commander, Emanuel Hamelin, was an extremely competent sailor who'd go on to become an admiral. He was also much more able to maintain harmony on his ship, the *Naturaliste*, than Baudin could on *Géographe* – although this may have been due to the reduced number of civilian scientists Hamelin had on his hands. Nevertheless, the ships missed so many rendezvous that for much of the expedition they travelled alone. Baudin eventually sent Hamelin back to France with the more undesirable elements of the expedition, replacing *Naturaliste* with a smaller shallow-draught vessel, *Casuarina*. Again his problem was communication with its capable but young and ambitious officer, Louis de Freycinet. Freycinet missed rendezvous even while in plain sight of Baudin or failed to report discoveries that might affect or change Baudin's plans.

Having two ships didn't necessarily mean things were safer, either. *Géographe* and *Naturaliste* collided at least twice; Cook and Furneaux's ships collected each other while in the Antarctic ice. If the ships were sailing together, each with a competent commander, they could still end up wrecked together, as both La Pérouse and Flinders would discover.

The crew's health

The most important element of any expedition was the crew and its well-being. A competent and healthy crew was essential for continual maintenance and management of an exploration vessel over extended periods. Add to this the ongoing vigilance required of those brave enough to venture within the uncharted coral labyrinths of the Great Barrier Reef or among icebergs in polar regions, and only a fit crew was likely to return.

During this period the biggest destroyer of mariners wasn't shipwreck or even war. It was disease, predominantly scurvy. The symptoms of this disease, caused by lack of Vitamin C, were initially spontaneous weariness, heaviness and stiffness, and difficulty breathing. Then would come rotting gums, nasal bleeding, dizzy spells and swelling, and finally gangrene and haemorrhaging, externally or internally, as blood vessels burst. A seaman's life was fraught with danger at the best of times, but to perform tasks high aloft in a pitching sea and in other extreme conditions with any stage of scurvy would accelerate the likelihood of an early death by accident before the disease itself produced the same result. Over the twenty years of naval warfare during the Napoleonic Wars, of 103 660 dead suffered by the British Navy, over 80 per cent were from disease and accident with only around 6 per cent killed in action, and the difference from wrecks, fires and explosions.[2]

No-one knew then what caused scurvy. It wasn't until the twentieth century that anyone did. Sadly, even if the cause wasn't known, effective cures to the disease had been in use long before Cook's voyages. As early as 1593, English Admiral Sir Richard Hawkins had noticed the efficacy of oranges and lemons in a sailor's diet. Still, appalling losses weren't checked. The ultimate expression of the problem was Anson's voyage of 1740. Anson left England with a squadron of eight ships and 1955 men. Four years later only one ship, *Centurion*, with fewer than 200 men returned. They'd lost 997 from scurvy, 300 from typhus/dysentery; only four were killed in action; and the remainder in shipwreck and starvation. To add to the tragedy, in 1747 a young naval surgeon, James Lind, proved in a controlled experiment on board *Salisbury* that patients fed oranges and lemons recovered quickest from scurvy, whereas those on other diets took longer to recover or didn't. Although Lind published his findings in 1753, these continued to be ignored in preference to the solutions of less practicable but more powerful lobbyists. Lind was a great friend of Joseph Banks and in fact joined Banks on the substitute expedition to Iceland following Banks's withdrawal from Cook's second voyage. This may explain why Banks was aware of the restorative quality of lemon juice and had several forms of citrus juice and extract with him, the most effective version being one mixed with alcohol, which acted as a preservative. He resorted to this private stock when once afflicted by scurvy's symptoms on *Endeavour*, with excellent results.

Ironically, for all Cook's fame as a campaigner in the war against scurvy, here it could be argued that he did more harm than good. Cook's trip may well have been intended as an anti-scurvy experiment, given the number of different remedies available to him. Cook's scurvy preventatives consisted largely of sauerkraut, salted cabbage, essence of spruce, and malt. Cook also had lemon rob, an extract obtained by boiling lemon juice, but the extraction process tended to destroy most of the Vitamin C, leaving the lemon rob relatively ineffective. Sauerkraut did have some Vitamin C content and therefore some value. Essence of spruce and malt had none at all. Yet no man on *Endeavour* died of scurvy. The Admiralty therefore generally continued to support the use of malt, spruce essence or sauerkraut, but the Sick and Hurt Board had already stated in 1767 that oranges and lemons were ineffective against scurvy.

Given the length of Cook's voyages, no-one can blame the Admiralty for thinking that he had some miraculous cure. But it's more likely that Cook's crews survived because he was fanatical about ensuring his ship and crew remained as clean and as dry as possible – he required his crews to keep a dry, clean set of clothes. He also made sure his crew had fresh food whenever available. Sauerkraut would've supplemented these measures in terms of preventing scurvy, but would not have worked alone, as Banks's own experience shows.

A desire to constantly replenish water and fresh food was nothing special, and most voyagers spent much of their time procuring fresh supplies from foreign ports and keeping their staple salted fare for hard times. It was only common sense. Cook's regular haunts for food, wood and water were Tahiti, New Zealand and Van Diemen's Land, places where his crews also could be rested. Cook also introduced a three-watch system that ensured eight hours' sleep to crew in lower latitudes.[3] This system, together with the emphasis on cleanliness and dryness, meant that Cook's men were less wet, cold and tired than many of their contemporaries and so used up their reserves of Vitamin C more slowly. The Admiralty provisions themselves did little to replace these reserves. With arrogance typical of the times, the Admiralty assumed that it was Cook's, that is, the Admiralty's, provisions, rather than common sense, that had eradicated scurvy from Cook's voyages. The result was that the same relatively useless remedies were prescribed for Admiralty trips for another thirty-five years, and that other countries, particularly France, copied these 'remedies' for their own crews with far more disastrous results. All on Baudin's expedition believed that mineral water was also a remedy. Interestingly, one of the reasons for the relatively low mortality – thirty-two deaths – on Arthur Phillip's First Fleet after an eight-month voyage was Phillip's purchase of Brazilian oranges at Rio de Janeiro, specifically in line with Lind's remedies, in 1787.

Change came only twenty-five years after Cook's first voyage when, in 1795, Sir Gilbert Blane of the Sick and Hurt Board finally persuaded the Admiralty to order the issue of lemon juice and sugar on its ships, not quite fifty years after Lind's findings and 200 years after Hawkins's. On the barren northern Australian coastline, where fresh food and water could be extremely scarce, even these measures were insufficient. Flinders had a supply of lemon juice with him, but only as a thirst quencher. He would suffer from scurvy himself in these regions.

In accordance with Admiralty views of the time, Bligh went on to use Cook's remedies on *Bounty*, with identical results to Cook, for much the same, totally misunderstood reasons. Having directly benefited from Cook's tuition as master of *Resolution* on Cook's last voyage, it stands to reason that Bligh attempted to press upon *Bounty*'s crew similar requirements for their well-being. Vancouver did likewise and had no death from scurvy and only one from disease on the two ships under his command over more than four years. Yet where Cook executed his health regimen so well, this approach worked so poorly for Bligh and Vancouver. It seems perverse that history judges one as humane and the others as tyrants for the ways they handled their crew when each appears to have applied similar principles with equally keen regard for their crew. Each saw scurvy in any crew as a sign of a negligent commander.

The French experience demonstrated the disastrous effect of relying on Cook's remedies. In many cases they were equally as successful in combating the disease but, again, probably more for the fresh food they took on board during the voyage than for the remedies supplied by their government. French commander Bruny d'Entrecasteaux died of scurvy towards the end of his expedition in 1793 while searching for La Pérouse, and at least eighteen cases existed on board his ships at that stage, with more developing.

Places such as Timor and Batavia (Jakarta) were the closest available resupply points to Australia. They were also unsewered cesspits of dysentery, typhus, cholera and malaria. Fresh water seemed almost impossible to obtain there. Despite Cook's eradication of scurvy, he lost over half of his crew and passengers on *Endeavour* from disease contracted at Batavia on their return voyage to Britain. Thirty years later Baudin called at Timor on his outward, rather than homeward, voyage. The introduction to his crews of these diseases – many, like dysentery, mortal by themselves – so early in their three-year voyage accelerated the onset of scurvy. Dysentery, for instance, absorbs the body's Vitamin C reserves quickly. Baudin was the only navigator unfortunate enough to have dysentery on board his ships so soon. Not surprisingly, scurvy started to appear in the crew much earlier than normal, and he still had two further years of voyaging to contend with. Baudin did well to avoid major mortality, but this was largely because of the proximity of Port Jackson for fresh food, rest and recuperation. Due to the legacies of his earlier visit to Timor, by the time Baudin reached Port Jackson he had fewer than ten healthy men, from a total of seventy-five, with whom to sail the *Géographe*.

Later surveyors were to have the benefit of these earlier travails, and were able to combat scurvy more effectively. By the time of Phillip Parker King's voyages in 1818, not only was lemon juice a settled part of the daily issue, but advances in food preservation in tins allowed an important variation to a salted meat diet. Healthier crews meant that order could be maintained over longer periods, allowing for longer expeditions. Scurvy still remained a real risk. It plagued Robert Scott's men in his last and tragic Antarctic expedition in 1912.

The crew's morale

Crew management was the other obvious essential to these long and arduous voyages. Even assuming a physically fit crew, the brilliance of a commander's orders or of an expedition's achievements would come to nothing unless within the ship there was a balance between respect for the commander's decisions and the harsh naval discipline of the day, which didn't allow these decisions to be questioned. Most sailors, captain and crew alike, preferred to ship with old familiar faces so they could rely more on past shared experiences and less on naval discipline in order to achieve the required level of shipboard harmony. On routine naval service in familiar waters, where the duty was somewhat repetitious, this level of familiarity and ease could often be the norm. But the navigators weren't dealing with the norm. They were dealing with the new. They wouldn't, as a rule, have the benefit of a crew seasoned to the dangers and rigours they would face – most of these were unknown. Many crewmen of the *Endeavour*, who could draw on their experience from the two voyages of the *Dolphin* to the Pacific under Byron and Wallis, were an exception to this. As previously stated, this is one reason for the success of Cook's voyages, and it also contributed to *Endeavour's* reputation as a happy ship.

The navigators had an additional human element to deal with. This was the presence of civilians on board in the form of botanists, astronomers, gardeners, artists and other 'savants', who could – and did – cause huge discontent. The French, especially Baudin, had significant difficulties on this score. The British learnt earlier than the French to avoid these arrangements, meaning that by the early eighteenth century all 'savants' were also naval officers or else present in exceptional circumstances and clearly under naval command. The naturalist Thomas Huxley, for example, began as a naval surgeon's mate.

An efficient and happy crew could still be upset by unforeseen or unusual influences, which naval discipline wasn't used to dealing with. The best example of this is the mutiny on Bligh's *Bounty* during its voyage from Tahiti. This voyage had a number of characteristics that made it exceptional. While a significant factor would've been the distractions of the sexual pleasures of Tahiti, many other ships visited Tahiti without mutiny. There were other key contributing elements to the mutiny, including extremely cramped living conditions, caused by the breadfruit stowage, the absence of the usual contingent of marines on board to maintain order – even Cook had a detachment on *Endeavour* – and a demonstrated lack of confidence by Bligh in his few experienced officers. Bligh's rude and provocative language didn't help, nor did his parsimonious management of the purchase and distribution of rations. However, in many ways Bligh's behaviour was typical of naval officers of his time. He may have felt he was doing no more than copying his idol, Cook, who on the third voyage with Bligh had demonstrated this sort of behaviour.

The mantle of command and Cook

Clearly Cook could inspire and command. *Endeavour* was a happy ship, and many of Cook's crew from that ship followed him on future voyages. Little is known of Cook's personality other than from the accounts of those who accompanied him on the voyages. His wife destroyed all his correspondence after his death. The only other record of Cook is in the private journals of his officers. As these always had to be handed in to the Admiralty at the conclusion of a voyage, they usually contained little criticism of the leader, as this could prejudice the writer's future career prospects.

Cook was interested in new ideas. He was able to adapt himself to and master the challenges of self-education, sailing, navigation, surveying and chartmaking, despite coming from a poor family and receiving limited education. Having risen through the ranks himself he also knew more of the common sailor's hardships than most naval officers of his time and rarely forgot these. However, he had his limits. The considerable stresses of his previous voyages had strained his composure as a leader by the time of his third and last voyage. It showed in the excessive punishment he administered both to his crew and to the pilfering natives of the Pacific. It also showed in the way he navigated his ship on that voyage.

Cook had demonstrated a degree of daring with navigation of his ship in the past, but there's a fine line between courage and recklessness. His achievements during his first two voyages, whether among Queensland's coral or Antarctica's ice, were courageous. On his third voyage, as we've seen, the Admiralty had set Cook a massive task – one never to be resolved in sailing ships – find the North-West Passage linking the Pacific and Atlantic oceans. Cook was by then almost fifty; he was tiring and less than well. He'd also turned down an offer of comfortable semi-retirement, a sinecure as head of Greenwich Naval Hospital, in order to take command, but only after quiet persuasion from the First Lord of the Admiralty, Sandwich, and others. By the second year of this third voyage, the ice was defeating Cook, and it was becoming clear that this expedition wouldn't be marked with the successes of the others.

As this trip wore on, Cook's mind appeared to become affected. His officers began to lose confidence in his judgment. He'd horrified them by sailing fast downwind off the Alaskan coastline in heavy fog when they could hear breakers sounding in the gloom ahead. They hove to and anchored, and when the fog lifted they found themselves surrounded by rock, with breakers only a few hundred yards ahead.[4]

Cook had always maintained fairly tough discipline with his crews, nevertheless his concern for their welfare had always been high. Now this aspect of his leadership, too, deteriorated. On this third voyage he implemented some dubious regimens. In the Arctic Circle, in experimental pursuit of fresh food over salted, Cook ordered his men to eat walrus meat and put those who refused on a diet of ship's biscuit only. He didn't have the same success that he'd had with the sauerkraut almost ten years previously. He was forced to restore salted meat rations after men began to starve.[5] He would've come close to mutiny when he cut the grog ration entirely after the crew refused to drink a substitute beer he'd manufactured from a recent purchase of Hawaiian sugar. Order only returned with reinstatement of the grog ration.[6] Cook was well intentioned in his concern for his crew, but so dogmatic and obdurate in enforcing his will on others that he failed to realise the harm he was doing. He was lucky that his capable junior officers could handle these excesses. In fact they frequently referred among themselves to his rages as a *heiva*, a word they'd learnt in Tahiti, which was a loud dance with much stamping.

Cook's ill-judged behaviour wasn't confined to the crew. At Moorea, not quite eighteen months before his death, following the theft of two small ship's goats, his treatment of a community of natives bordered on barbaric. He ordered a heavy-handed reprisal. In one area alone, a huge number of canoes and paddles were destroyed by his men, depriving the natives of their chief means of sustenance. The injustice of the act wasn't lost on his subordinates.[7] Thieving in these communities wasn't unusual. It's difficult to fathom why it provoked such a response from one so familiar with Pacific Islanders as Cook had become by this time. But worse was to come. Not long after this incident a native passenger on board *Resolution* was discovered with stolen goods. Cook ordered the barber to shave the man's head, a common form of punishment and ridicule, but also to cut off both his ears. Luckily a junior officer intervened, but not before part of an earlobe had been cut away.[8]

It was a similar, almost reckless attitude that ultimately led to Cook's death. Seeking recuperation after their unsuccessful efforts to find the North-West Passage, Cook's ships discovered the Hawaiian Islands. There they were treated as gods and enjoyed a festive period of recovery, but only because their arrival fortuitously coincided with the local spiritual calendar. Their departure was similarly celebrated, probably all the more so due to the depletion of the islands' scarce resources in lavish entertainment of these gods. However, Cook's unexpected return to the islands shortly after to repair a split foremast definitely didn't comply with any religious occasion. The Hawaiians were far less friendly and failed to follow the usual submissive pattern of other Pacific communities. Following what had become an increasingly violent number of smaller incidents, the theft of a large cutter was to Cook the last straw. His response was to attempt to kidnap the Hawaiian king and hold him as hostage for the cutter's return, something he'd often done with the Tahitians and the Maoris. However, while he was going about this, news reached the chief that Cook's men had killed a minor chief in a skirmish across the bay. This provoked sufficient aggression from the Hawaiians to cause Cook to retreat from the chief's village with his small group, and to fire his musket at them. Cook fired only buckshot and intended to hurt rather than kill. However, the buckshot simply bounced off the woven armour of the threatening warriors. They rushed forwards with heightened confidence, assuming all the British muskets to be equally harmless, which of course they weren't. As Cook only had a small marine detachment with him, they were quickly overwhelmed. Even as they went down under the onslaught, Cook appeared detached from reality. Some observers thought that he could have run and escaped but, with almost ambling disregard, he walked to the water and was cut down with his hands protecting the back of his head.

The officer responsible for firing the first shots that day had a reputation for following Cook's orders with fanatical dedication. Cook could rely on diligence from this obstinate character. He may in later life have been haunted by the fact that his was the first act of British violence in the conflict that culminated in Cook's death that day. The officer was Bligh.

$\mathcal{W}$ILLIAM BLIGH, THE SURVIVOR OF COOK'S SCHOOL

William Bligh was arguably the most persevering and longest-living of the British navigators. Although he didn't chart long lengths of the Australian coastline, through Bligh's tutelage we see the development of a young Matthew Flinders. We then follow Flinders' short life, the characters he meets and the extraordinary rivalry with Nicolas Baudin's French expedition, both in execution of the voyages and publication of the results.

Bligh is an absorbing character who, luckily for us, was able to cause sufficient stir during his lifetime that much was written about, and by, him that survives to this day. That more is written against him than for him says more about the forcefulness of his character than of the truth of the facts related.

Early days, and *Resolution*

Like Cook, Bligh commenced naval life almost at the bottom as a 'volunteer' seaman, which meant he could mess with the midshipmen, the lowliest of the officer class, although rated only as able seaman. Despite the fact that his name appears as early as 1762, at the age of eight, on naval records as captain's servant, this position was only for a few months. It would in any case have involved little actual service at sea. At the time, in order to sit formal lieutenant's examinations, youngsters had to first prove that they'd put in six years of time at sea, evidenced by entry on a ship's register. Fudging the records by entry at an early age was an accepted practice to evade this rule, and was usually only available to those with strong naval connections.

Through his early career Bligh had to continually volunteer as an able seaman both to obtain a berth on ships and to obtain sufficient time at sea to sit his lieutenant's exam, a classic sign of one with no connections whatever. By May 1776 he passed his exams and was promoted to lieutenant. Bligh was almost immediately appointed sailing master on *Resolution* for Cook's third voyage. At the age of only twenty-two, this was an impressive appointment for someone with no connections – no 'interest' – within the navy. It isn't really clear which particular interest had 'pulled strings' for him. Possibly he was at that time already known to Joseph Banks. Perhaps someone with interest had noticed that he was eminently capable, as by then he'd also acquired some charting skills. It could be that his aptitude to learn these skills reminded someone of Cook. The actual reasons remain a mystery, and Bligh's own journal from that voyage – it seems he kept one – has never been positively identified.

The appointment to *Resolution* stood Bligh in good stead. Not only did he serve under the illustrious Cook but this also put him in touch with the circle of highly placed individuals interested in both Cook and the voyage – particularly Banks, who was shortly to be elected president of the Royal Society. Banks would remain a major patron of Bligh during his career.

That Bligh needed interest or influence at all to advance in the navy may have put an enormous chip on his shoulder throughout his career. What interest did for others less capable than Bligh – but not for him – was to be made clear to him again and again, particularly in the aftermath of the *Bounty* mutiny.

Bligh appears to have found favour with Cook, itself an achievement given the change others had noticed in Cook during his final voyage. Bligh assisted in much of the chartmaking and took over this task following Cook's death. Nevertheless none of this was acknowledged upon publication of the charts; the credit was given to Henry Roberts, a more junior master's mate, who was subsequently promoted to lieutenant. Roberts had sailed with Cook on his second voyage and had done all the charting copywork for both Cook and Bligh. Given that public recognition for any navigator only came through publication of his work, Bligh never forgot this perceived slight. Ironically, Bligh's inability to share credit for charting work with his own subordinates caused friction some years later.

Bligh and *Bounty*

Having been paid off from the *Resolution* on return to England, Bligh married. His bride was Elizabeth Betham, whom he'd met on the Isle of Man while on leave. He also met two other families there with close ties to his future wife, the Christians and the Heywoods, both families with excellent connections in the navy. His wife's uncle, Duncan Campbell, was a trader, plantation-owner and ship-owner from the West Indies. He was related to Vice-Admiral John Campbell. Bligh now had as much 'interest' as he needed. Both as husband to Elizabeth, and ultimately father to their six daughters and their twin sons, who sadly died at birth, he was loving and loyal.

Bligh saw the end of the American War of Independence as a junior lieutenant in a ship of the line. He returned to the merchant shipping command offered by Duncan Campbell and remained there for the next five years. This would've been far more attractive than the struggle ashore in naval employ during peacetime on a meagre lieutenant's half-pay. On the last few voyages in Campbell's employ, Bligh agreed to take as a volunteer foremast hand a young Fletcher Christian as a favour to one of the families close to his wife.

Bligh's next appointment was to the *Bounty*, a naval vessel. *Bounty*'s task was to bring breadfruit from Tahiti to the West Indies. On first consideration, the appointment appears a little surprising for someone of Bligh's background, but it's easy to see how it came about – the machinations of Campbell and Banks had combined. Campbell would have been well known to Joseph Banks, given his trading links with the West Indies and elsewhere. Campbell was also the original owner of the *Bounty*, then named *Bethia*. Banks would also have heard about Bligh as a result of Bligh's involvement in Cook's last voyage.

Together with Lord Hawkesbury, Banks was the chief promoter of the *Bounty* voyage. He intended the voyage to demonstrate the use of science in conjunction with the navy to the nation's trade and commerce by proving that plants from one climate could be transferred to others for the economic benefit of Britain and its colonies. In the wake of the American War of Independence, the cost of buying food from the now economically free, liberal – and monopolistic – United States had risen dramatically. Breadfruit would provide a much cheaper food source for the black slaves of the West Indian sugar plantation owners than was currently available, keeping their sugar production costs, and therefore the sugar price paid by Britain, low. This was Enlightenment in action, the pursuit of scientific knowledge as a means of serving the commercial and social interests of State. The State, represented by Banks's great friend Lord Hawkesbury of the Committee for Trade, was only too happy to support the voyage, and not for reasons of Enlightenment alone. The voyage also supported Hawkesbury's ongoing enforcement of the British Navigation Acts against the American colonies. These Acts effectively prevented American ships from trading with the West Indies. By providing planters with a cheap food substitute for their slaves, the breadfruit voyage would also reduce the need for the planters to trade with the American colonies in defiance of the Navigation Acts.

The voyage of the *Bounty*, its famous mutiny, and Fletcher Christian's involvement in it, is well known, and won't be dwelt on in detail here, except to say that the evidence doesn't justify any conclusion that Bligh was any more tyrannical in his behaviour than even Cook, other than perhaps in his tendency to abusive tirades. Certainly others, such as Vancouver, flogged their crews mercilessly (Bligh didn't), ignored their health (Bligh and Vancouver enforced a scurvy regimen similar to Cook's with the same successful results) and welfare. Bligh even took on board *Bounty* a blind fiddle player to encourage the men to 'dance and skylark' for exercise.

What's more relevant is that the voyage heralded an amazing navigational feat. Bligh, cast adrift by the mutineers – crammed into the *Bounty*'s 23-foot open launch with eighteen others – with no charts and only a compass and a sextant, sailed over 5700 kilometres from Tonga to Timor in forty-three days in April–June 1789. He even made a rough chart of his voyage. This required the negotiation of the uncharted Australian Barrier Reef and the elusive Torres Strait. Bligh had to rely entirely on his memory of his charts to set his course to Torres Strait. Having no watch, he trained his men to count out seconds as a marked logline was trailed from their stern to calculate their speed of travel. Only one man died on the voyage, killed by natives when landing for water. David Nelson, the Banks-appointed botanist, died shortly after their arrival.

Upon Bligh's return to Britain on 14 March 1790, less than eleven months after departing *Bounty*, the ill-fated frigate *Pandora* was sent by a vengeful Admiralty to find the *Bounty* mutineers. Bligh in the meantime was a public sensation. He was quick to capitalise on this. He published a narrative of his experiences, was acquitted on court martial for the loss of his ship, and was appointed post-captain by the end of that year.

Bligh and the second breadfruit voyage, on *Providence*

A second breadfruit expedition was planned, in the two ships *Providence* and *Assistant*. Why go to the trouble and expense of repeating the *Bounty* voyage? Because of the breadfruit, of course, and the huge political cachet for Banks and Hawkesbury that would go with its success. For them, the *Bounty* mutiny had been a disaster. *Providence*'s voyage, promoted by Banks and various West Indian plantation-owners, no doubt Duncan Campbell among them, was needed to demonstrate the importance of Enlightenment in action. Bligh, now the toast of London, was the obvious man for the second breadfruit voyage, and was commissioned to command it in April 1792.

This trip was every bit as uneventful as the *Bounty*'s had been fraught with incident; the breadfruit was again collected from Tahiti and safely delivered to Jamaica. Bligh crewed his ships with relations and other trusted supporters. He had with him a contingent of marines, which had been absent from the *Bounty*. He also agreed to take someone in the midshipmen's berth on the recommendation of Captain Thomas Pasley. Although Pasley was the uncle of one of the young mutineers from the *Bounty*, Midshipman Peter Heywood, he was also extremely powerful in the navy. Pasley had recommended one Matthew Flinders.

Bligh, like Cook, was concerned to educate his 'young gentlemen' in the skills of seamanship, including navigation and chartmaking, and it's clear that young Flinders had the benefit of this. Certainly Flinders appears to have thought himself a little overused by Bligh in this regard by the end of the voyage, having assisted in the preparation of the voyage's many charts, for which Bligh and his subordinate George Holwell took all the credit on publication. Bligh's memories of his own mistreatment for the benefit of Roberts in similar circumstances following Cook's third voyage would've reminded him, one would think, of the importance of fair treatment here, if Flinders' complaint was justified. Perhaps it wasn't. Flinders was only seventeen at the start of the voyage and probably did much of the background work in the chart production, rather than the primary work. He was likely to have been far too young and inexperienced to merit greater mention in the publication of these charts. As we shall see, Flinders was always well aware – sometimes more than those around him – of his own worth.

Another aspect of Bligh's second breadfruit voyage that went smoothly was the hydrography. He was able to add four sets of eleven new charts (forty-four in total) to the Admiralty collection. These included Bligh's Islands – actually the Fiji Group, but clearly an indication of Bligh's own ego at work – and Banks' Islands to the north of the New Hebrides. There were also charts of the Friendly Isles – the Tongan Group – and, most relevant for Australia, a northern passage through the Torres Strait, Bligh's Entrance – again the ego at work. This is still an important entrance to that waterway today.

By the time of Bligh's return from successful delivery of the breadfruit in 1793, the luckless *Pandora*, returning from Tahiti with fourteen *Bounty* mutineers, had been wrecked on the Great Barrier Reef. Christian, eight other mutineers, the *Bounty* and nineteen Tahitians had disappeared to Carteret's elusive Pitcairn Island, their whereabouts to remain unknown until the next century. Ten mutineers survived *Pandora*'s sinking, then faced a far worse ordeal than Bligh's open boat trip when they made a similar voyage to Timor as prisoners in the *Pandora*'s open boats. Their subsequent return to England and 1792 trial in the absence of the main prosecution witness, Bligh, created a huge public outcry. The French Revolution was by then in full swing and the rights of man were in hot debate across the Channel, with strong discontent in London as well. The senior surviving mutineer, Midshipman Peter Heywood, was well connected within the navy –

his uncle, Thomas Pasley, was captain of the powerful ship of the line, *Bellerophon*. Fletcher Christian's family was similarly extremely influential in legal circles and supported Heywood. Bligh didn't have powerful connections but, worse than that, as he was still at sea on *Providence*, he was *not there*.

Bligh's demise

Of the ten survivors of the *Pandora* wreck, four, including the blind fiddler Michael Byrne, were acquitted of mutiny. They were never mutineers in the first place. There had simply been no room for them in Bligh's launch and so they'd remained on *Bounty*. However, on the orders of *Pandora*'s captain, Edwards, they'd been locked up and treated like mutineers ever since the arrival of *Pandora* at Tahiti; they almost drowned in their chains when *Pandora* sank. The other six mutineers were sentenced to hang. Clemency and connections, however, prevailed, so that three of these, including Heywood, received pardons. The others were executed on 29 September 1792. In Bligh's absence Edward Christian, Fletcher's brother and a prominent barrister, had published a defence to his forever absent younger brother's actions to which Bligh was unable to reply until his return in the *Providence*. If Bligh *had* attended the trial with the evidence he'd collected in Tahiti on the second voyage in *Providence*, Heywood probably would've been hanged.

By the time of Bligh's return from his voyage, the outcry from the mutineers' trial had yet to die down. The Admiralty was somewhat embarrassed by his presence and was less interested in his forty-four charts than they might have been. Although a successful commander would usually be able to use his influence to promote the interests of his men, clearly this wasn't the case with Bligh; at this stage his support tended to be more a kiss of death for their chances for promotion. Furthermore, no publisher was interested in any narrative of Bligh's second voyage, despite the popularity of his earlier accounts of the *Bounty* mutiny. Even some twelve years later this remained the case. The most Bligh could salvage from the situation was a reward for successful delivery of the breadfruit, voted by the West Indian plantation-owners.

While young Flinders was still on board the *Providence* on its return voyage, Pasley wrote to him on 7 August 1793 to warn him: 'Your captain will meet a very hard reception. He has dam'd himself.'[1] The ambitious young Flinders subsequently sought alternative patronage for his advancement, largely from the only man to benefit from Bligh's efforts, Banks.

Bligh's second mutiny

Bligh's naval career was able to slowly repair itself, largely because career opportunities became available with the commencement of Britain's war with the French and Spanish in 1794. But it seems that mutiny was something Bligh could never quite escape. In command of *Director*, ship of the line, Bligh faced the second mutiny of his life while at the Nore, one of the main anchorages for Britain's North Sea Fleet. At least this time he wasn't alone. In May 1797, following a relatively bloodless mutiny at Spithead of the Channel Fleet the previous month, the entire North Sea Fleet stationed there mutinied. Although the mutineers ejected most other ship's captains and officers from their ships directly, they allowed Bligh to retain command of *Director* for a week after his most hated subordinates had been sent ashore. Once ashore Bligh was consulted closely by Evan Nepean, secretary of the Home Office, and Admiral Duncan about how to deal with the mutiny. When it finally broke and the mutineers were forced to surrender, Bligh actively worked to obtain pardons for many of his men. Although his lieutenants had reported twenty-nine mutineers on *Director*, on his return to the ship Bligh wished only to make examples of ten of these. When some hundred officers in sixteen ships were removed from their commands by Lord Howe in the inquiry that followed, Bligh wasn't one of them. Given his reputation from the *Bounty*, his record would have been more closely examined than most but he doesn't appear to have been found wanting. His demonisation by history would appear to have been brought about by the influence of the Christian and Heywood factions, rather than by fact itself.

Bligh's career, having faltered briefly, continued. As commander of *Director* at the Battle of Camperdown in October 1797, he was responsible for the capture of the opposing Dutch admiral's flagship, *Vryheid*. He was to command *Glatton* at the Battle of Copenhagen in 1800 under Horatio Nelson and led the third line of British vessels in the attack. Despite these successes, Bligh clearly still felt insecure with some factions in the Admiralty. After the battle he actually asked Nelson for what could only be described as a written reference to Admiral the Earl St Vincent. Nelson obliged, writing in his dispatch that:

[Bligh] has desired my testimony to his good conduct, which although perfectly unnecessary, I cannot refuse: his behaviour on this occasion can reap no additional credit from my testimony ... the moment the action had ceased ... I sent for him ... to thank him for his support.[2]

Seen as a courageous captain by most, and a tenacious one at that, his name came up as a replacement for Captain King as governor of the troubled colony of New South Wales in 1806. Not surprisingly, his advocate for this was Banks.

Even then Bligh had recently been in the news again at a court martial instigated against him by a subordinate from when he was commander of *Warrior* in February 1805. One Lieutenant Frazier accused Bligh of 'gross insult and ill treatment', and 'tyrannical, oppressive and unofficer-like behaviour contrary to the Articles of War', including Bligh calling him 'rascal, scoundrel and shaking his fist in his face'. Although acquitted by the court, Bligh was cautioned in his use of 'bad language' with his officers, in its own way a quiet backhander in an institution as tough and rough as the Royal Navy at that time.

Bligh was a competent, zealous and determined officer who was prepared to take tough decisions, as a leader should. He could obviously recognise when these qualities might be lacking in his subordinates. Bligh's problem was the way he dealt with their deficiencies, usually by telling them so, usually publicly, bluntly and in a rage or with insulting language. This behaviour was probably akin to one of Cook's *heivas*. Once the rage passed, Bligh was only too keen, equally as publicly, to let bygones be bygones, but by then it was too late. Pride was irreparably damaged, especially with young, inexperienced subordinates. On a closely packed, uncomfortable and competitive ship, this sort of behaviour could be destabilising, as witnessed in extreme form on the *Bounty*.

Bligh's third mutiny

Banks must have felt that Bligh was just the sort of governor needed for the colony of New South Wales in 1806. The colony was at this stage suffering economically from the monopolistic control imposed by ex-officers of the Rum Corps and leading merchants. It would take a tough individual to re-establish authority. At fifty-two, Bligh was getting old for active sea duty and otherwise faced the prospect of life ashore as a post-captain on half-pay. There were clear benefits to him in accepting the post. But it was scarcely in Bligh's character to bother with the velvet gloves, so when pitched against the greed and corruption of the New South Wales Corps and the merchants and landowners of Sydney, things came to a head. In the Rum Rebellion of 1808 Bligh faced the third mutiny of his life, the humiliation of being dragged from an upstairs room – reportedly from under a bed – of Government House by the rebellious soldiers of the New South Wales Corps and, ultimately, return to Britain and quiet retirement with his family as a rear-admiral. Eventually order was established in New South Wales with the dispatch of Lachlan Macquarie, and his supporting regiment of soldiers, as the first non-naval governor of New South Wales.

Bligh's beneficiary

Yet again, the man who would benefit more from Bligh's efforts to preserve the colony was the same one who'd supported the colony from its conception, and the same man who'd benefited from Bligh's *Providence* voyage – Joseph Banks. Men like Bligh were of great value to Banks. Without them Banks couldn't execute any of his projects or prove his own worth as effectively. Even by the time of *Providence*'s return, Banks was well entrenched as president of the Royal Society, de facto scientific adviser to the government and, more importantly, King George. Any promoter of a scheme with any commercial value and scientific, geographical or mechanical content requiring government support would seek an entrée from Banks into the nebulous world of Home Office, the Privy Council Committee for Trade and the Admiralty.

Banks's capability as a political operator is even more defined when one compares what he got from the breadfruit voyages against their ultimate result. Banks acquired further power, prestige and position as a man who could originate and execute projects for the benefit of the State. The West Indies acquired a plant for cultivation in Jamaica's freshly rejuvenated Botanical Gardens. By the time it became commercially available it was refused by the local inhabitants in favour of their local tapioca diet, and was therefore effectively a failure. By then, however, other projects had taken its place.

CHAPTER NINE

THE NEW BLOOD, MATTHEW FLINDERS

Bligh's second breadfruit voyage in *Providence* appeared to be of little scientific value in the long run. The forty-four charts remained unpublished, as did the official narrative. The breadfruit was never used. What the voyage did was to introduce Matthew Flinders to the 'bloodline' of navigators.

Flinders' life and his voyages, as well as those of his French rival, Nicolas Baudin, exemplify the stresses and successes of the navigators. These two characters demonstrate the remarkable qualities that navigators required for survival and success in the turbulent world at the turn of the eighteenth century.

Early days, with 'interest' on *Providence*

Born into a Lincolnshire family of small means, Flinders' initial connection or 'interest' with the navy was through his cousin, Henrietta, governess to the family of Captain Thomas Pasley – *Bounty* mutineer Peter Heywood's uncle and strong supporter. Flinders was appointed to Pasley's ship *Scipio* in 1790, then he transferred with Pasley to *Bellerophon* shortly after. As we've seen, on Pasley's recommendation, in 1791 Bligh agreed to take the seventeen-year-old 'young gentleman' on *Providence* for the second breadfruit voyage.

Although initially happy to earn Bligh's favour, things don't appear to have been harmonious between Bligh and the young Flinders. For example, at first Flinders was proud of the fact that Bligh entrusted him with the care of the ship's three chronometers, but given Bligh's reputation for bad temper and strong language, one wonders if Flinders took the brunt of a tongue-lashing when whoever was responsible – Bligh merely referred to the culprits as 'my young pupils' – knocked them over during Bligh's astronomical observations at Tahiti in June 1792. The instruments apparently never quite worked the same afterwards.[1]

Flinders might also have earnt Bligh's criticism for wholly different reasons. Bligh, like Cook, was critical of the sexual pleasures available from the natives of Tahiti, would certainly not have submitted to them himself, and not only discouraged seamen from partaking of them but wouldn't allow women on the ships. More than anyone else, Bligh understood their disruptive effect from *Bounty*. Cook had even punished sailors for consorting if afflicted with venereal disease, for fear of infecting the otherwise clean inhabitants – such restrictions would've had little effect with the passage of some twenty years of ships' visits since. Flinders, like other young men of the ship, freely indulged in the sexual attractions of Tahiti. Geoffrey Ingleton's research discovered evidence of Flinders being treated for 'venereals' twice in the three-month period in which the ships were at Tahiti.[2] It's quite plausible that Flinders suffered other such bouts that went unrecorded and probably untreated, as sexually transmitted disease wasn't something one made public. This wasn't to avoid embarrassment. The reason for likely secrecy was that if discovered, treatment would be prescribed by the ship's surgeon. Treatments were expensive and were docked from one's pay – each cost the equivalent of two weeks' pay for someone of Flinders' rank. Also, Flinders would have been aware of Bligh's strict views on the subject and the likelihood of incurring his commander's disapproval if his liaisons became public.

Although clearly not recommended by Bligh for immediate promotion at the end of the voyage in 1793 – not that other deserving midshipmen on board were either – Flinders may not have been so upset. By then Bligh himself was so unpopular back in London following the trial and execution of the *Bounty* mutineers in his absence that many didn't try to associate too closely with him. As we've seen, Bligh had 'dam'd himself'.

Even if he didn't obtain a promotion, Flinders would've gained valuable experience in navigation, seamanship and rudimentary surveying and mapping from this voyage. He also struck up an important friendship with James Wiles, the botanist appointed to the trip by Banks to care for the breadfruit during the voyage to the West Indies. Flinders even lent Wiles the best part of £40, as the Admiralty had been tardy in paying Wiles. Whether or not Flinders knew this at the time, this would be one of the best investments of his life. Upon his return to England he met Banks for the first time when recovering the debt from Banks at Wiles's direction.

The Glorious First of June, and lucky *Reliance*

Following *Providence*'s arrival back in England, Flinders returned to a position in Pasley's *Bellerophon* in October 1793 and saw action on that ship as part of Admiral Thomas Graves' squadron in Lord Howe's fleet victory – known as the Glorious First of June, 1794 – fought in the Atlantic some 600 kilometres west of the Isle of Ushant, at the tip of Brittany. This gained Flinders some prize money, £10, and a recommendation for promotion. Again with the influential Pasley's support, Flinders was appointed to his luckiest ship, *Reliance*, bound for New South Wales.

Reliance is almost unknown in the history books, although many individuals brought together sailing on her would become famous. These included Captain – and later Governor – John Hunter and Lieutenant John Shortland, after whom New Guinea's Shortland Islands are named. George Bass, the surgeon of *Reliance*, would be mentioned in history almost as often as Flinders.

Hunter already had strong connections with New South Wales. He'd journeyed there as captain of *Sirius* with Captain Arthur Phillip's First Fleet in 1787. *Sirius* became stranded on Norfolk Island and wrecked while landing a small contingent of colonists there in 1790.

The Dutch merchant vessel *Waakzamheyd* was hired by the colony to take Hunter and most of his crew home via Batavia and the Cape. En route she passed Carteret's Admiralty Islands, relevant to the failure of a future French expedition.

Hunter reached Plymouth in 1792. He was now returning to the colony to relieve the tiring Arthur Phillip as governor, taking with him both *Supply*, under command of one Captain Kent, and the newly purchased *Reliance*.

Selection of the commander of *Reliance* shows how 'interest' worked in the navy of the day. *Reliance* was to be commanded by Nathanial Portlock, but he declined at the last minute. Portlock was highly qualified. He'd been master's mate on *Discovery* on Cook's third voyage and made further exploration on the north-west coast of America as a member of the King George's Sound Company in 1785–88. He was commander of *Assistant*, the consort vessel to *Providence*, on the second breadfruit voyage with Bligh. Following Portlock's change of heart, the command of *Reliance* was then offered – through Portlock and Hunter – to Henry Waterhouse. Waterhouse had been recently a lieutenant on the *Bellerophon* and the appointment was no doubt made with some influence from Pasley. Waterhouse was also well known to Hunter, having served as midshipman on *Sirius* and lieutenant on *Waakzamheyd* with him. Flinders obviously was on good terms with Waterhouse from their days together on *Bellerophon*, and was appointed master's mate on *Reliance* as a result of this last-minute reshuffle.

Flinders was developing 'interest' most effectively through Pasley and must have been liked by the officers from *Bellerophon*. Further evidence of this was the addition as 'young gentleman' to *Reliance* of Samuel Flinders, Matthew's twelve-year-old brother who was, as we'll see, perhaps more deserving of recognition than history now – or possibly his ambitious elder brother then – would allow.

Waterhouse wasn't impressed by *Reliance*. In a letter to Phillip he described her as 'so complete a tub, which the first lee shore we are caught upon will prove.'[3] His words would be prophetic.

Reliance and *Supply* left Portsmouth in 1795, reaching New South Wales in August. *Supply* only just got there, her American timbers unsuitable for service in warm waters. *Reliance* limped in shortly after. She was so 'complete a tub' that she wouldn't be fit for further duty without considerable repair. Although on station at Port Jackson until 1800, *Reliance* was only useful for the occasional trip to the Cape, to Norfolk Island or following extensive refit. Although she spent a total of roughly three years in Sydney Cove undergoing refit, Flinders was away from her for no more than six months. But what a productive six months it was.

The Socratic Bass, and the two *Tom Thumbs*

Governor Hunter realised that the newly settled colony was in need of a surveyed coastline. Cook had capitalised on a similar need in Newfoundland over thirty years before. Hunter, himself an experienced navigator and surveyor, had conducted surveys of Port Jackson, Botany Bay and Broken Bay when with the First Fleet. Like Governor Graves of Newfoundland had done for Cook, Hunter was only too keen to encourage further coastal survey by Flinders, who was keen to oblige. Although gifted with an inquisitive, tenacious and mathematical mind, Flinders was short on experience in commanding ships and giving orders, rather than receiving them. What Flinders lacked in this he made up with audacity. No doubt Hunter also taught him as much about surveying as he could.

Flinders' partner in his early trips, George Bass, was even more audacious. As Bass's services as surgeon weren't essential to *Reliance* during the refit in port, unlike Flinders he was left idle and restless. In fact Bass must be one of the most short-lived, but certainly the most remarkable, of Flinders' associates, and it's likely that some of his energy rubbed off on Flinders. Perhaps they complemented each other, Flinders' analytical nature and potential as a navigator coming to the fore when inspired by Bass's adventurous spirit.

George Bass was a terrific character. He came from Lincolnshire, like Flinders, and there's some evidence that the two were distantly related, which may explain the close friendship that arose between them. Certainly they grew up not far from each other, one apprenticed to a local doctor and the other a son to one. Clearly a highly intelligent man, Bass had trained as a surgeon before succumbing to a desire to go to sea. His family had been able to buy for him a share in a merchant ship that was subsequently wrecked, leaving him little alternative but to join the navy as a surgeon to satisfy his restless spirit – and some pressing bills. Bass was six feet tall, vigorous and strong, of 'well informed mind and active disposition',[4] and with a keen sense of humour. Bass seemed to take on hard tasks for the enjoyment of conquering the challenge. Flinders referred to him as the 'penetrating Bass'[5] and said of him, 'my mind has often called you its Socrates'.[6] Hunter described him as 'of much ability in various ways out of the lien of his profession'.[7]

Bass's sense of whimsy is apparent from various descriptive passages in his diaries. He said that the dying-song of swans 'exactly resembled the creaking of a rusty ale-house sign on a windy day', and remarked that the entrance to Twofold Bay was notable for 'a red point on the south side of the peculiar bluish hue of a drunkard's nose'.[8]

During his study of surgery in London, Bass had the benefit of contact with the influential botanist William Curtis and the great anatomist John Hunter – not to be confused with *Reliance*'s John Hunter, future admiral and governor of New South Wales. Bass was also well read in natural history, certainly not areas of general knowledge to a naval ship's surgeon. He would put this knowledge to good use on a later voyage with Flinders.

Bass's first exploratory trip with Flinders was made between 26 October and 3 November 1795 in Bass's tiny 8-foot dinghy, *Tom Thumb*, with Bass's servant, Martin, exploring the Georges River. Bass's report on the soil ultimately led to further survey by Hunter and the establishment of Bankstown.

A second *Tom Thumb*, probably larger than the first, was built, ostensibly to replace *Reliance*'s jolly-boat – or small work boat. Flinders, Bass and Martin set off in it between 25 March and 2 April 1796 to explore further south of Botany Bay, where a river entrance was rumoured to exist. They were carried further south than intended by the current and experienced their first adversities in the new country, being unable to find shelter to anchor for the night. They also ran short of water. Bass swam ashore through the surf with a cask to replenish supplies, and the small craft capsized in the surf when venturing in to pick him up. With food, clothes and muskets thoroughly drenched, they reloaded the little craft and pushed out again through the now high surf, and didn't clear it for hours. Bass would've been exhausted by now. He'd been labouring in the sun, naked, throughout the day – from when he'd first gone over the side with the empty water cask and tried to swim it back full. He was extremely sunburnt and possibly suffering from exposure. They tried to land for the night on one of the Five Islands off Port Kembla, but the wind was too strong. A cold night was passed offshore in the boat. They knew they had to land again, to find water, dry provisions and repair an oar that had broken during the rescue ordeal.

Finally they ventured inshore near what is now Port Kembla. They were soon surrounded by a group of Aborigines, who led them up a rivulet – the ocean outlet from Lake Illawarra. On a sandy beach they repaired the broken oar, dried their powder, clothes and provisions, and collected water. They became rather concerned that the Aborigines were leading them further upstream and away from escape should things turn at all ugly. The Aborigines became excited and aggressive when Bass began to clean the sand-filled and rusty muskets. Bass was also by then wearing a red jacket that appeared to remind the Aborigines of red-coated British soldiers, as they uttered a similar-sounding expression, 'soja', when they saw his coat. The tense situation was defused with a shiny pair of scissors, which were employed cutting their observers' beards and hair, much to the amusement of the onlookers.

After they left this group and returned to the coast, they were hit by a 'southerly buster' on their return voyage. They only just survived this gale by taking shelter in a tiny cove at present-day Wattamolla in Sydney's Royal National Park. They named this life-saving haven Providential Cove. Six kilometres further north, they discovered their original goal, Port Hacking, which they spent a couple of days examining before returning to Port Jackson. A sketch map produced from the voyage was eventually published in London in 1799. The caption at the bottom would have stirred Flinders. Finally his name had appeared in print.

Only two months after the return of the second *Tom Thumb* voyage, Bass was again off, this time to conquer the Blue Mountains, the as yet uncrossed mountain barrier between Port Jackson and the unknown Australian inland. Bass had his own scaling irons made to allow him and his small party to climb cliffs. Armed with ropes and provisions, they headed inland in June 1796. Fifteen days later he and his party emerged defeated and exhausted. Even though they'd pushed further up the Grose River than any others, probably as far as the Wentworth Creek junction, they were unsuccessful nevertheless. The mountains would remain uncrossed for a further seventeen years.

In 1797 Bass's discoveries, along with reports of coal near modern-day Newcastle by Lieutenant Shortland, prompted Governor Hunter to write to Banks reminding him of the need for a coastal survey. *Reliance*, with Bass and Flinders aboard, then made a trip to the Cape of Good Hope to obtain livestock for the still hungry colony. The stormy weather on the return trip damaged *Reliance* so badly that she was immediately laid over for repairs on her return. The repairs would take up all of Flinders' time, again leaving Bass bored and restless.

News of a maritime tragedy led Bass back to Providential Cove and fresh discoveries. In February 1797, en route from India to Port Jackson, the trading vessel *Sydney Cove* had suffered structural damage from battling several severe storms, and had developed a serious leak off Van Diemen's Land. To save the ship's passengers, crew and valuable cargo, Captain Guy Hamilton beached her on Preservation Island, a small island in the Furneaux Group at the entrance to Bass Strait. Hamilton had sent eighteen men in the ship's longboat north to Port Jackson for help. Unfortunately the longboat was wrecked on the mainland at Point Hicks, leaving the men an overland march of more than 500 kilometres. Only three survived, to be rescued by a surprised fishing party near Bass's Providential Cove on 17 May 1797.

The survivors burnt coal they found on the beach to keep warm.

The enterprising Bass returned on 5 August 1797 with two of the survivors in Governor Hunter's open 28-foot 7-inch whaleboat, also discovering the first of New South Wales's many coal seams in the cliff face near present-day Wollongong. He named the spot where he found the coal Coal Cliff. The coal was a significant find; three bags were collected for Governor Hunter, and specimens were sent on to Sir Joseph Banks in London.

He also discovered the remains of two of the *Sydney Cove* castaways, who'd been killed by the local Aboriginal people, apparently over the castaways' insistence on taking their fish. In fact, from Bass's discussion with the local Aborigines – Bass could speak some of the local dialect – it transpired that their killer was Dilba, one of the men who'd guided them up the Lake Illawarra estuary in *Tom Thumb* the previous year. Perhaps their apprehension at that time had been justified.

After this, with Flinders still preoccupied with *Reliance*, Bass undertook his most audacious expedition. He asked Hunter for permission to explore as far down the south coast as was safe to go. Hunter agreed and provided this unusual naval surgeon with his whaleboat and six volunteers, including one worthy of note, *Reliance*'s bosun, John Thistle.

Bass's Strait

On 3 December 1797 Bass and his crew headed south in the open whaleboat with six weeks' provisions. His voyage would show reasonably clearly, but not conclusively, that a strait separated Van Diemen's Land and the mainland. He would also discover Western Port and its outlying islands – complete with a group of hungry and dispirited runaway convicts – off what Hunter would name Wilsons Promontory, apparently after a relatively unknown London merchant.

Strong westerly winds and an easterly current had convinced Bass that, contrary to the conclusions of Furneaux and Cook, a passage did exist here – we now know it as Bass Strait. It was a hard voyage, particularly once they reached the rough waters of the strait, where an eight-day storm with high seas and increasing leaks in the open boat would have been exhausting. By New Year 1798 they were off unknown coastlines and on 5 January they discovered Phillip Island and Western Port. By this stage Bass judged from both the tidal flows and the strong southwesterly swell that a strait separated Van Diemen's Land from New South Wales, although the only way to be sure would be to attempt a circumnavigation of Van Diemen's Land.

Shortage of provisions prevented further coasting westward to prove this conclusively, and denied them discovery of the massive Port Phillip Bay. They headed back to Port Jackson. Bass made contact again with the runaway convicts. The convicts, with several others, had stolen a cutter in Port Jackson and sailed south, attracted by the remains of the wrecked *Sydney Cove* further south, from whose timbers they'd hoped to make a seaworthy, ocean-going vessel for escape. However, their colleagues had abandoned them one night and sailed away, never to be heard of again. There was only sufficient room in the boat on the return voyage for two of the convicts, the oldest and weakest. The other five had to be left to their fate on the mainland with what food Bass could spare and a musket. They were never seen again.

The party returned to Port Jackson on 25 February, after voyaging for fifteen weeks and living off seals, birds and fish. Bass was treated as a hero. The whaleboat would enjoy cult status thereafter; visitors to Port Jackson in the future reported it as preserved for posterity.

Bass's sketch map of the area, although a useful first impression, would require reworking. Lack of any reliable timekeeper, as well as the fact that the observations had been taken from a small, open boat, tossing in Bass Strait, meant that his chart references were inaccurate by almost 16 kilometres. That he, a ship's surgeon, had made one at all in such conditions and with minimal charting experience was scarcely credible.

Norfolk, Flinders' navigating debut

Flinders' real navigating debut came in late 1798, in *Norfolk*. At 25 tons and just 33 feet in length, *Norfolk* was a small sloop that had been constructed on Norfolk Island for carrying dispatches to Port Jackson.[9] According to some, she was the revamped and decked longboat of the long gone *Sirius*, rebuilt on Norfolk Island and sailed to Port Jackson in June 1798.[10]

Sailing south on *Norfolk* in November 1798 with Bass and Roger Simpson, who was from the whaler *Nautilus* – but again *without* a timekeeper to accurately fix longitude – Flinders charted Twofold Bay in New South Wales and a month later mapped the Tamar River in what Hunter would call Port Dalrymple (accessing Launceston). From there they headed west to the Hunter Group and then down the west coast of Tasmania, although sufficiently offshore to miss the entrance to Macquarie Harbour entirely. With Flinders in command, Bass could devote himself to the pursuit of natural history. He collected many specimens of plant and animal species, some of which were preserved and shipped back to Sir Joseph Banks. *Norfolk* also explored the Derwent River, and Bass was probably one of the first Europeans to have significant contact with the Aboriginal people there.

The return across up the east coast proved conclusively the existence of Bass Strait – as Flinders insisted, to his credit, it be named after that audacious spirit – and that Van Diemen's Land was but an island.

The *Norfolk* voyage was a milestone for Flinders. By 12 January 1799 they were back at Port Jackson. Hunter dispatched Flinders' chart of Van Diemen's Land to the Admiralty on 15 August 1799. It was printed the following year, the first published chart of Van Diemen's Island. In London Flinders' achievements attracted favourable attention in high circles, just as Cook's Newfoundland work had for him thirty-five years before.

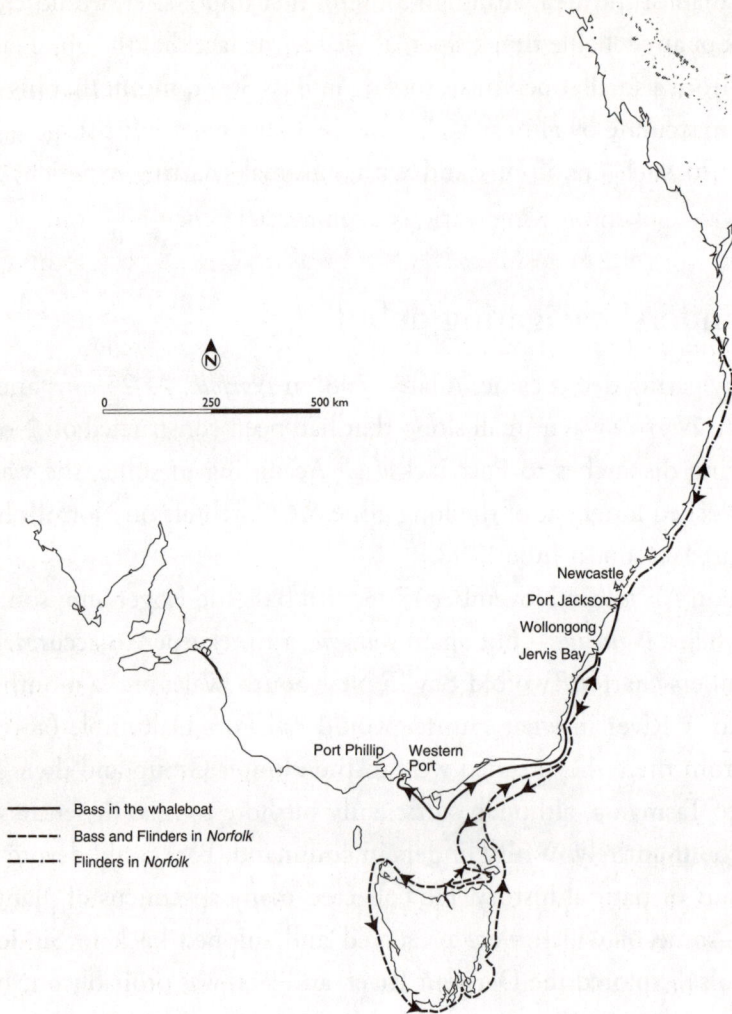

Bass's whaleboat voyage and the *Norfolk* voyages

The end of Bass

The *Norfolk* voyage was Bass's last exploratory trip. Bored with life in Port Jackson and apparently fatigued from his various adventures to date, he asked for a discharge from the navy on the grounds of ill-health and left for Macau in May 1799 with Messrs Simpson and Bishop of the merchant vessel *Nautilus*. Bass was happy to take risks for reward, particularly material reward. He'd seen the merchants in Port Jackson become very rich in the undersupplied and monopolised colony. He'd reportedly joined in the speculation, importing sheep for private sale on *Reliance* during its voyage to the Cape of Good Hope for livestock in 1797. He later suggested to Governor King various commercial schemes – one for the import of spirits from Batavia in late 1802 to break the already blossoming rum trade, the other to commence a New Zealand fishery in early 1803. King, however, sought clearance from London before moving forward on either.

With Bishop, Bass would float a company and buy a new brig, the *Venus*, for trading with the colonies. Flinders no doubt saw him again after *Reliance* returned to Portsmouth in 1800, as Bass married Elizabeth – the sister of Henry Waterhouse, who'd commanded *Reliance*. Elizabeth saw Bass for the last time in January 1801, when the *Venus* sailed from England with a cargo of trade goods for New South Wales.

A market glut made Bass's cargo unsaleable in New South Wales, so he was forced to run trips to Tahiti and Hawaii, supplying Port Jackson with pork and beef, to make ends meet. Hunter's replacement, Governor King, wouldn't take Bass's trade goods off his hands, even at a 50 per cent discount. To make matters worse for Bass, during the voyage from Tahiti, his business partner, Bishop, responsible for sailing *Venus,* began to show signs of mental imbalance, leaving management of both the ship and the trading business to fall on Bass's shoulders.

Bass also appears to have felt a little neglected by the government, probably over their hesitancy to help him shift his trade goods, and refused to show to King his journals and sketches from his pork voyages to Tahiti and Hawaii.[11]

There was, however, a huge black market for trade goods in the Spanish colonies on the west coast of South America, as Spanish ships and traders had a legal monopoly on trade there. This monopoly, together with the remoteness of these colonies from Europe and attacks on shipping by British privateers – even before Spain officially went to war with England in 1804 – led to exorbitant prices and short supply for European goods. Even though the profit potential was enormous, to sell to this market meant risking prison and confiscation of property by the Spanish.

This sort of risk wouldn't stop Bass, who wished to bring his wife out to New

South Wales. In February 1803 he wrote to his brother-in-law, Waterhouse:

> ... in a few hours I sail again on another pork voyage [i.e. to Tahiti], but it combines circumstances of a different nature also ... Speak not of South America to anyone out of your family, for there is treason in the very name.[12]

Governor King had also given Bass what he described as a 'very diplomatic-looking certificate'[13] intended to support him visiting South American ports to obtain supplies for Port Jackson 'without any view to private commerce or any other view whatsoever'.[14]

Bass sailed from Port Jackson on 5 February 1803. He was never seen again.

Later in 1803 the brig *Harrington* reportedly heard that the Spanish had caught Bass in a Chilean port, confiscated *Venus* and any cargo, and dispatched Bass and his mate, Robert Scott, as prisoners to the silver mines.[15] The Spanish would have been suspicious of any ship from Port Jackson, which they regarded as a nest for British privateers. Any cargo of contraband trade goods on *Venus* was now extremely valuable, and any excuse for confiscation would have been attractive to the poorly paid Spanish officials. Considering the simmering state of peace between England and Spain at the time, it's conceivable that imprisonment of Bass would have been discreetly managed to avoid diplomatic tension. The war that followed no doubt sealed this result.

There's always been debate about *Harrington*'s report of the fate of Bass and whether this was confused with another *Venus* that was seized in Peru for trading in contraband. But this didn't occur until 1807, four years after Bass had disappeared.

Then there's the *Harrington* and the sources of the report itself. The story was effectively only hearsay and not even made known to Bass's relatives until 1811, eight years after he'd disappeared, by someone who'd been living in Chile in 1808 and may have confused the two ships. *Harrington* was commanded by a Captain Campbell, who'd caused a potential international incident between Britain and Spain after returning to New South Wales from contraband trade in Chile and Peru in 1805 with two captured Spanish prizes. Even though this was a major newsworthy item for Port Jackson, there was never any mention of Bass. As

Bowden points out in his biography of Bass, it seems strange that such a mention would've been overlooked in the colony, unless it'd never actually been made.[16]

There were reports that Bass was alive in Lima as late as 1817 but these were never substantiated. Maybe Bass was simply lost at sea, like so many others in those days. He'd planned to visit southern New Zealand en route to the Pacific Islands, in order to strip the hulk of a wrecked ship there of iron for sale to the islanders. In those cold and dangerous waters *Venus* may have suffered a fate that even Bass's audacity couldn't overcome. It was a sad end for such a keen spirit as George Bass to die friendless and in chains in a Spanish silver mine or in the cold southern oceans.

Norfolk heads north

After Bass's departure for Macau in May 1799, Flinders took the *Norfolk* for a second voyage, north along the New South Wales coast, this time with a chronometer. His objective was to more accurately chart this coastline and to locate major rivers. The British believed that such a large continent as New Holland must either be drained to the ocean by large rivers, in which case there would be a seaway inland for sail access and further exploration, or else contain a large inland sea, possibly accessible by one large outlet. Flinders hadn't been able to locate any large river outlets further south than Port Hacking. Neither had anyone else. So, if there were no big rivers draining the land, the run-off had to end up in an inland sea, or possibly a north–south strait.

As he took *Norfolk* north, Flinders checked carefully for river outlets past Newcastle and the mouth of the Hunter River. With him were his brother, Samuel Flinders, a local Aboriginal identity named Bongaree, and eight seamen. Between 8 July and 20 August 1799 they explored as far north as Hervey Bay. Despite the close survey, Flinders managed to miss all the important northern rivers in New South Wales ... at least eight of them. Although Flinders could be anything from 3 to 25 kilometres offshore, he was obviously not used to the deceptive nature of Australian river mouths, often narrow inlets hidden by long surf-pounded and scrub-ridden sandbars or sheltered mangroves. It was a challenge that would taunt subsequent surveyors and explorers, including Nicolas Baudin and, twenty years later, Phillip Parker King. It would also mean that much land-based exploration was required over the next two decades to find an answer to what became termed 'the riddle of the rivers'. This riddle might have been solved by exploration from sea mouths of major rivers, had they been located. For the moment, however, the British believed that a north–south strait dividing the east and west of Australia was more likely to explain the unusual water drainage of the continent. They simply had to find it, and explore the inner continent, before anyone else intent on settlement did, particularly the French.

The Flinders brothers returned to England on *Reliance*, arriving at Portsmouth in August 1800. Hunter had ordered the ship home as he was concerned that, with her deteriorating hull, she wouldn't withstand the arduous voyage back if he waited longer. He was right. On arrival *Reliance* was immediately paid off and dismantled.

CHAPTER TEN

$\mathcal{F}$LINDERS' BOLD DASH

Flinders' return home set the stage for his most ambitious dash. Like Bass, seeing many others make their fortunes in the colony from speculation had given him food for thought. He weighed up his options for bettering his lot in life. To make serious money in the navy of the day he needed to capture, not destroy, enemy shipping, from which he'd be entitled to a generous share of the prize money. For this he needed both promotion to commander and, from there, to post-captain. This would hopefully lead to the command of a fast frigate or a ship of the line. Promotion was probably the most important thing for all of this, and it required interest he didn't at that stage have.

In a recently unearthed letter Flinders wrote to Bass in February 1800, Flinders appears to have been less than enthusiastic about further naval surveys. He not only offered himself as a business partner to Bass, but he also made it clear that he was dissatisfied with naval service in general, and the idea of conducting further surveys in Australia in particular, without promotion or some extraordinary attraction.[1]

Yet unlike the infinitely more impulsive Bass, he wasn't prepared to throw away his gains in the navy to date. In the event, Flinders decided that his best avenue for advancement was another expedition of survey and exploration, of which he would be likely to receive command, and post-captaincy, only if he was the main promoter. But where? The remaining uncharted coasts of New Holland were along its often inaccessible northern and south-eastern coastlines. Why would anyone be interested in these? Also, in a time of war when available ships for this sort of thing were few and far between, how would he get hold of the necessary ships? Clearly he would need to leverage some influence. In this his strategies were masterful.

Flinders knew that Governor Hunter had been impressed by his efforts along the coastlines of New Holland and would say so to his superiors and to Hunter's close acquaintance, Sir Joseph Banks. Banks knew of Flinders from Hunter's reports and may also have recalled meeting him as a young midshipman when paying him Wiles's £40 debt in 1793.

The Admiralty and Banks must have already thought highly of his charts to date. In March 1800 they'd dispatched a small survey vessel, the *Lady Nelson*, to Port Jackson specifically for Flinders' use – Chapter Eleven will cover the *Lady Nelson* in more detail. Flinders may also have known that Evan Nepean, the secretary to the Home Office, had his hands full with the difficulties of running State affairs for a country at war with France. He would delegate the organisation of such an expedition to Banks, president of the Royal Society, close friend of Lord Hawkesbury and the ubiquitous *ex officio* minister for science and the colony of New South Wales.

Importantly, Flinders may also have known, or been told, that the French were preparing an expedition under Nicolas Baudin of two ships – the most lavish and well equipped to date – to explore the same coastlines. The French had applied in June 1800 to the Admiralty for a passport of free passage for Baudin's ships, despite the state of war between the two countries, in order to do this. This application had been referred to Banks. On 6 September 1800 Flinders wrote a letter to Banks, proposing a further investigation of the New Holland coasts. It was a bold move, and the letter itself a *tour de force*. It draws upon almost every known argument which would have found favour with Banks, and one must wonder if Banks himself hadn't first suggested to Flinders its form; some believe that Banks had encouraged and supported James Matra in this way, almost twenty years before, when Matra proposed the reasons for settlement of New South Wales. In addition to the unstated but massively compelling political desire to pre-empt the French in exploring the Australian coastline, Flinders' letter covered all the right arguments.

Flinders appealed to Banks's interest in trade and 'Enlightenment'. He posited the likelihood of a north–south strait between the Gulf of Carpentaria and the Great Australian Bight, and that '… advantages of such a strait to any settlements in Van Diemen's Land or New South Wales by the very expeditious communication with India, seem almost incalculable'.[2] He went on to suggest that exploration of Torres Strait and southern New Guinea could be achieved in short order, and that to the extent that cannibals on these coasts made life difficult, the expedition could recharge itself in Bligh's or the Feejee Islands (Fiji). The latter suggestion is interesting, as he would have known from Bligh's experiences, as well as his own on *Providence*, that the Fiji Group was inhabited by cannibals. He recommended that two vessels rather than one should be employed, of which one, at least, should be bigger than the *Lady Nelson*.

He then expanded on this theme in the letter. The additional size of the vessels Flinders requested allowed him to recommend that:

… a person or persons be accommodated who should examine into the natural productions of this wonderful country, for surely what has been found is materially different from all others, and the mineralogical branch would probably not be the least interesting … [3]

The theme of the letter – trade and the use of science in enhancing this – would have been music to the ears of Enlightened 'dilettanti', especially Banks, who had in a similar role sailed with Cook and who promoted and profited from enlightened schemes such as the breadfruit voyages of Bligh. He was also aware of the discoveries of coal around Sydney.

The letter naturally mentioned that if Flinders should be considered for command of the expedition, he wouldn't refuse. Then it closed with a beautiful exercise in understatement, quietly claiming a right of kinship from the Lincolnshire fens from which Flinders had come, and over which Banks, as wealthy landowner and occasional high sheriff of Lincolnshire, prevailed. As a deft finishing touch,

Flinders enclosed a packet of seedlings, flowers and plants from New South Wales.

Flinders was also starting to look at a more settled personal life with Ann Chappelle, a family friend for many years. On 25 September 1800 he wrote to Ann:

> You see that I make everything subservient to business. Indeed, my dearest friend, this time seems to be a very critical period of my life. I have been long absent, have services abroad that were not expected, but which seem to be thought a good deal of. I have more and greater friends than before, and this seems to be the moment that their exertions may be the most serviceable to me. I may now perhaps make a bold dash forward, or remain a poor lieu-tenant all my life.[4]

Definitely a bold dash forward for a 26-year-old – Cook was forty when appointed to *Endeavour* – but one that even today would strike a chord for many people in a range of callings.

The letter produced a reply from the great man, requesting that Flinders call on him *at any time*, meaning that as far as things were concerned, *Flinders was there*. This happened on 16 November 1800. Only five days later the first of many Admiralty directions sanctioned by Earl Spencer, the First Lord of the Admiralty, was sent to numerous departments, contractors and suppliers, selecting the ship *Xenephon*, to be renamed *Investigator*, for the expedition.

The high point, appointment to *Investigator*

On 19 January 1801 Flinders was appointed to the *Investigator* and he was pro-moted to the rank of commander on 16 February 1801. His first lieutenant would be another Lincolnshire man, Robert Fowler, only a year younger, and his second lieutenant, Samuel Flinders. Flinders' young cousin John Franklin, aged fifteen, was appointed midshipman. Flinders also extracted a promise from Earl Spencer that he'd be promoted to the coveted rank of post-captain following his return from New South Wales in *Investigator*. Flinders was lucky to have set the wheels in motion when he did. In February 1801 a change in government saw Earl

Spencer replaced by John Jarvis, Earl St Vincent, as First Lord of the Admiralty, who would develop quite a different view of Flinders.

Six civilians, largely Banks-appointed 'savants', would accompany Flinders. They were botanist Robert Brown, botanist's assistant Peter Good, mineralogist John Allen, landscape painter William Westall, natural history painter Ferdinand Bauer, and astronomer John Crosley.

Flinders had gained enormous experience during his time with *Reliance* and used this to huge advantage in securing, if not creating, his appointment to *Investigator*. However, when one compares his experience with Cook or even Bligh, it's an extraordinary appointment. Flinders had been a lieutenant for only two years, much of which had been spent in port. Flinders' charts to date were only running surveys, noting only major headlands rather than detail of the land in between, and he certainly missed many of the major rivers.

Nevertheless, being wartime and with fleet movements planned in the Channel for the coming spring, there was probably no appropriate alternative commander available. Bligh was too senior, and Vancouver was dead. All the other experienced officers from 'Cook's school' were dead, infirm or retired, or in service elsewhere. Flinders' experience, Banks's support and, importantly, the British concern with possible fallout from the French expedition under Baudin, weighed heavily in favour of the expedition with Flinders as commander. The British feared that the French desired to settle a base in Australia or nearby. Now that it was known that Van Diemen's Land was separate from the mainland claimed by Britain, following the discovery of Bass Strait, the little island was an easy target for settlement by the French, and the British had no tangible grounds for objection. A base like this would cause havoc with the trading activities of the British East India Company, disrupt trade with – and the survival of – the colony of New South Wales and the more general aspirations of the British in the region. The current French base at Île de France (Mauritius) on the trade route between India and the Cape of Good Hope was a good example of the threat. French privateers operating out of Mauritius caused huge losses to British shipping and trade throughout the Napoleonic Wars.

Flinders was scrupulously thorough in his planning. He visited old Alexander Dalrymple while in London. Dalrymple now headed the Hydrographic Office of the Admiralty, an office only five years old. Dalrymple, the original and jilted promoter of the *Endeavour* voyage, would only live another eight years. Nevertheless he was important, not only as the ultimate recipient of all hydrographical work conducted by any future expedition, but also due to his string of connections within the establishment.

Flinders again demonstrated Machiavellian aplomb. At his first meeting with Dalrymple he presented him with a survey – which Flinders had conducted while on the *Norfolk* – of a previously unknown port, suitable for settlement, and named in his honour by Hunter as Port Dalrymple.

The old cartographer must have been impressed. Three of Flinders' maps and an accompanying pamphlet of observations were published by the hydrographer's subcontracted printers, Aaron Arrowsmith, on 20 February 1801. Flinders was *it*. As the newly promoted young master and commander of *Investigator*, and with a clear track to post-captaincy, his luck from *Reliance* had held yet again.

CHAPTER ELEVEN

LITTLE *LADY NELSON*, THE FORGOTTEN DISCOVERER OF PORT PHILLIP

At the time of Flinders' departure from New South Wales in *Reliance* in 1800, it was common knowledge both in London and Port Jackson that the colony was poorly endowed with survey vessels. Bass had discovered Bass Strait in a 28-foot whaleboat, and Bass and Flinders' *Norfolk* wasn't much bigger. Governor Hunter's earlier requests for a sensible survey vessel hadn't been lost on Joseph Banks, the éminence grise of the establishment on matters concerning the colony of New South Wales.

In May 1799 Banks heard of the development by the Transport Board of a special form of cutter with a shallow draught of only 6 feet and three sliding keels. The design had been the constant project of one Captain Schank for many years, with smaller vessels built for revenue and customs duties in England. One can only guess how he felt when, in 1799, the long arm of Joseph Banks was able to seize the most recently completed vessel of this sort from the Transport Board for the Admiralty's purposes. Banks's intention was to deliver it to Sydney for use by Flinders – in an unambiguous sign that Flinders was in favour even before he wrote his letter of 6 September. The vessel was tiny, just over 52 feet in length overall. She was the 60-ton *Lady Nelson*. She was nicknamed 'HM *Tinderbox*' because of her size, particularly when it became generally known that she was to be sailed all the way to New South Wales. For such a tiny vessel this was thought to be a mad, almost suicidal undertaking, particularly when fully loaded for the voyage her main deck was only 2 feet 9 inches above the water. Lieutenant James Grant, an acquaintance of Schank's and an extremely capable sailor, was chosen for this task.

Lady Nelson must be about the least recognised but one of the most deserving of the survey vessels that examined the Australian coastline. As her commander, Grant was to become the first to pass along and chart the northern coastline of Bass Strait. *Lady Nelson* would be the first ship to enter Port Phillip. She'd give twenty-five years of service until her destruction and the murder of her crew at the hands of pirates in 1825 at Babar Island off East Timor while obtaining supplies for a starving British colony on Melville Island, near present-day Darwin.[1]

Grant left England on 18 March 1800 under orders to deliver *Lady Nelson* to New South Wales for Flinders' command. The Admiralty didn't know that at this time Flinders was already on his way home from the colony in the failing *Reliance* and would certainly not be at Port Jackson when Grant arrived in December 1800. Fate ensured that sometime in late 1800 the two necessary elements for charting Australia's coastlines passed each other going in opposite directions between England and New South Wales – the surveyor, Flinders, going one way on *Reliance* and the required survey vessel, *Lady Nelson*, under Grant, going the other. This timing was lucky for Flinders. It meant that he could promote the *Investigator* scheme and ultimately take command of a much bigger vessel than *Lady Nelson*.

Grant, on the other hand, was less fortunate. Although a fine seaman, he was neither hydrographer nor surveyor, and these were the skills that Governor Hunter and his successor, King, wanted. Grant had only agreed to the risky task of sailing

Lady Nelson out because of the promise of the command of a much bigger ship, *Supply*, a large transport of size similar to, and past consort of, *Reliance*. Unknown to poor Grant, but presumably not to the Admiralty, *Supply* was so rotten she'd been condemned in August 1797 at Port Jackson.

When Grant left England, news of the discovery of Bass Strait had still to get there. Despite huge gales, and against the expectations of all, he reached the Cape of Good Hope in good form. By then news of Bass's discovery had spread, and he received further orders to sail through Bass Strait, taking particular notice of the coast and headlands visible on either side. Grant was the wrong man in the right place. Not a hydrographer, he'd unwittingly been landed with something for which he was unsuited. Chartmakers such as Flinders would instead have relished this opportunity to be the first to chart a new coastline, with a guarantee of publication and the kudos that went with all this.

Nevertheless Grant sailed the new coastline, compiling an eye-sketch from Cape Banks and Schank's Mount in the west to Wilsons Promontory at the eastern end of the Victorian coastline. He also noted the bay in which Port Phillip lies but didn't proceed further to investigate, thereby earning Governor King's quiet criticism upon his arrival in Port Jackson in December 1800. King also recognised the inadequacies of Grant's sketch and realised this coastline would need to be revisited by a far more skilled surveying hand. Grant's lot wouldn't have improved when he discovered that his future command had been condemned as a useless hulk.

King offered him the continuing command of *Lady Nelson* in the Colonial Service (i.e. doing what he wasn't trained for, charting and surveying, on the Colonial Service's much reduced pay). With little alternative, he accepted and was ordered back to Bass Strait to complete the survey with Ensign Barrallier, surveyor of the New South Wales Corps, and George Caley, the colony's botanist. Caley was an appointee of Banks, although an irascible one. Banks once said of Caley, 'had he been born a gentleman, he would have been shot long ago in a duel.'[2] John Murray, another individual whose name deserved more mention than it has received, was appointed as Grant's second-in-command.

Grant's chart of southern Victoria

On this voyage, between March and May 1801, Jervis Bay and Western Port were fully surveyed, but not Port Phillip, an omission that again earnt Grant the governor's displeasure. After a final voyage to survey the first 110 kilometres of the Hunter River and what is now Newcastle, Grant resigned from the Colonial Service in September 1801 because of his lack of surveying skills. Although he was the first to discover much of the Victorian coastline from Cape Banks to Cape Albany Otway, this has been forgotten over time, as has Grant. He left New South Wales in November 1801, as Flinders in *Investigator* was already approaching the west coast of Australia for the first time. Grant returned to Britain to serve the navy and was promoted to commander. He was pensioned off after being severely wounded in an heroic action off Holland in command of the cutter *Hawke* in early 1805. He died in 1833, promoted no further, aged sixty-one.

Now appointed commander of *Lady Nelson*, Murray was sent back by King in November 1801 to further survey Bass Strait, particularly the area west of Western Port. In early 1802 Murray would become the first European to enter and chart

Port Phillip Bay, following an initial reconnoitre by his first mate, William Bowen, in the ship's launch. Murray also charted King Island for the first time, although his work was appallingly incorrect.

While *Lady Nelson* was nosing its way along Bass Strait and Flinders was commencing his cruise in *Investigator*, the French were well advanced with their own revival of interest in Australia's coastlines. Nicolas Baudin had arrived on the Australian coastline in May 1801. What had rekindled the French interest in Australia after so many years?

CHAPTER TWELVE

THE FRENCH REAPPEAR

When we last saw the French, it was Bougainville making the sensible decision – based on his weakened crew and perishing rigging – to turn away from the Great Barrier Reef in 1767 and leave discovery of the Australian east coast to Cook three years later. What changes had occurred since to pave the way for Nicolas Baudin's well-equipped expedition on the Australian coastline, over thirty years later?

For the most part, French expeditions subsequent to Bougainville's had motives more obviously inspired by the pursuit of trade than science. Bougainville nevertheless became an important patron of the major French voyages that followed his own. Of course he was about the only leader of any of six French expeditions over the period of almost fifty years following his expedition to survive and return to France.

It isn't as though the French had only then begun to consider the Pacific. As early as 1740 Bouvet de Lozier had proposed to settle a trading base in the New Hebrides accessed by sailing south via Van Diemen's Land. The idea was to avoid having to negotiate with the Dutch and Spanish for permission to sail through their territories in the East Indies or South America, which would have made it difficult – if not impossible – to set up a competitive trading base. The French India Company, like the chartered trading companies of other nations, had a monopoly on the trading route and rejected this proposal. However, the advent of the Treaty of Paris had broken this monopoly, creating new opportunities.

Surville's 'near miss'

As we saw earlier, Bougainville's voyage had been prompted by the French defeat in the Seven Years' War and the consequent need for new trade routes and bases. The next expedition from France, led by Charles de Surville, was similarly motivated by the loss of Indian colonies to the British from the same war. Surville was engaged in a trading expedition to the East Indies and China. Following rumour of a continent 2400 kilometres west of Peru – Davis Land, which we met in Chapter Two – Surville stumbled across the Solomon Islands chain, seen by Bougainville the previous year. Eschewing contact with hostile inhabitants, and seeking wood, water and fresh food for his crew, he headed south. Finding nothing – although not far off the east Australian coast at the time – he struck southwest for Tasman's reported landmass, New Zealand, where he found wood and water but little food in early December 1769. Unknown to either of them, Cook was charting the same coastline further south at the same time. Cook then headed west to the Australian coastline. As we saw in Chapter Seven, he wisely accepted the counsel of his senior crew and decided not to head east because of the worn state of *Endeavour* and the onset of winter. With less luck, Surville headed east, seeking the rumoured great south land, which Cook had decided against looking for. Surville spent over two months looking until scurvy and refitting pressures resurfaced, forcing him to make for the Peruvian coast. Surville was drowned trying to land there on 7 April 1770, at the small town of Chilca, and the Spanish detained the rest of his crew.

Not for the first time, the French would be deprived of the future services of a commander and his now experienced crew, experience that the crew of *Dolphin* had demonstrated to be so valuable to Cook on *Endeavour*.

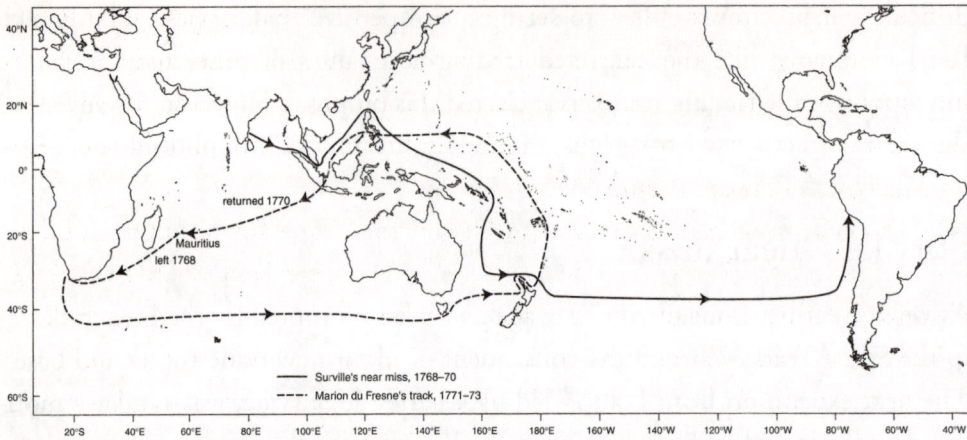

MAP 7

Surville's and Marion's tracks, 1768–73

Marion's innovation

Nicholas Thomas Marion Du Fresne's 1771–73 voyage came about because of Bougainville's obligation to return to Tahiti Ahu-toru, the native he'd taken with him to Paris in 1767. Marion, a landowner from Mauritius and a former sea captain, had found himself without employ following the demise of the French India Company after the Treaty of Paris. He floated a plan to both return Ahu-toru and search for the French equivalent of Dalrymple's *Terra Australis Incognita* – Gonneville Land. This was the alleged large landmass to which one J. B. Paulmier de Gonneville claimed to have been blown while sailing from the Cape of Good Hope to the East Indies in 1503. It was meant to be in the Indian Ocean, somewhere south-east of the Cape of Good Hope.

Marion was the first European in 130 years to head for Tahiti by its western rather than its eastern approaches, applying Bouvet's theory of voyaging in the high southern latitudes away from hostile trading competition from Dutch, Spanish or British colonies and relying on Tasman's reputed discovery of Van Diemen's Land as a resupply base. The voyage therefore had a character that was more than simply humanitarian, as any new route might lead to China and new trading possibilities. Trade in the Moluccas was planned for the return voyage.

In his strategy Marion was well ahead of his time – as Bouvet had been in 1740. The British were still focused on their colonies in the Americas, and Cook was only starting to plan a strategy similar to Marion's for his second voyage. But Marion wouldn't have Cook's luck. Ahu-toru died of smallpox shortly after leaving Mauritius. This didn't stop Marion.

Voyaging across the southern Indian Ocean from Cape Town, but not finding Gonneville Land, Marion's ships reached Van Diemen's Land in March 1772. They were the first European visitors since Tasman 130 years before.

Finding insufficient wood or water, Marion headed on to New Zealand, again in Tasman's wake, where the expedition stayed three months resting and refitting. Despite familiarity with the Maoris, he and sixteen others were killed and eaten after joining a Maori fishing party in the Bay of Islands, and eleven crew from a wood-collecting expedition suffered the same fate the next day. As much of Marion's planning hadn't been shared with his subordinates, the expedition returned through Guam and the Phillippines to Mauritius, a sad end to a well-conceived plan.

Yet again no coherent experience was passed on for future French expeditions, and another French commander was dead.

Kerguelen's fraud

The French remained keen to confirm or dispel the speculated existence of Gonneville Land, and sent Yves-Joseph de Kerguelen in January 1772 from Mauritius with this task. This was the first official French expedition dispatched for this purpose, with financial backing from the government and provision of two naval ships and crew. It was intended to restore France's reputation as a major player in maritime exploration. Kerguelen set out with two vessels from the Cape of Good Hope and sailed south-east. He discovered a green fogbound coastline in the high latitudes of the southern Indian Ocean that he was convinced was the western edge of *Terra Australis Incognita*. His glimpses from offshore of its green slopes were enough to convince him that here was France's potential future trading base. Despite the poor visibility and dangerous winds, François Alesno Compte de St Allouarn, commander of Kerguelen's second vessel, *Gros Venture*, ordered one of his boats to attempt a landing, which they managed, confirming that the land was desolate, windswept and icebound. What no-one realised at the time in the poor visibility was that the land was nothing but an island, as Cook later established, calling it Desolation Island. It's now known as Kerguelen Island.

Kerguelen almost immediately returned to Mauritius, never regaining contact with St Allouarn, and used his aristocratic connections within the navy to procure for himself another voyage to further explore this new discovery, effectively leaving St Allouarn and *Gros Venture* for dead.

St Allouarn, in the meantime, didn't follow Kerguelen. After he lost sight of Kerguelen, he headed east as originally instructed by the government, presuming erroneously that Kerguelen had done the same. He proceeded to West Australia's Cape Leeuwin then north to Shark Bay, charting as he went, in March 1772.

With Marion still in the east, this meant that two French expeditions were at the extremities of Australia at the same time, intent on exploration of parts that no Englishman had seen either at all – in the case of Van Diemen's Land – or since 1699, when Dampier was on the north-west coast. It's true that Cook had discovered the east coast of the mainland two years before, but that was largely due to luck and a chance route home, rather than planning. To the French, Australia was actually part of a strategy to supply new trade routes and bases without antagonising other powers.

St Allouarn sailed up the north-west coast, through what would later be named the Bonaparte Archipelago, to Melville Island near Darwin before heading to Batavia then back to Mauritius. Although he'd kept his distance offshore for most of this, it was a considerable reconnaissance for one small vessel in a continuous voyage off a desolate coastline. In later years Phillip Parker King would become famous for just such an effort. Despite the achievement, St Allouarn received no lasting fame. He died of illness soon after his return to Mauritius.

Kerguelen managed to persuade the French Government to fund another expedition to his new south land in October 1773, luckily departing before St Allouarn's return. Sadly Kerguelen's intentions for discovery were little more than a front for his own intended trading activities. Having once more located his desolate southern coastline, still shrouded in mist and foul winds, again he didn't attempt a landing and instead returned to Mauritius. This time questions were asked as to his failure, particularly now that reports had been received from St Allouarn's crew. Kerguelen returned to France to a court martial, substantiated accusations of fraud, and disgrace. Again, despite their labours, the French didn't have a surviving commander for future voyages.

Following these false starts, a period of eleven years – and the American War of Independence – was to follow before the French launched another expedition, under the influence of Charles Pierre Claret, Compte de Fleurieu, the French minister of Marine. Like Bougainville, Fleurieu had served in the Seven Years' War and he was also a keen hydrographer. He collaborated with eminent clockmaker Ferdinand Berthoud to develop a chronometer. Fleurieu believed in the importance of overseas trade and shipping in building up the strength of a nation. Although as part of his second voyage Cook had established that no great southern continent existed, what appeared from his third voyage were potentially profitable and as yet unexploited fur and whaling industries in the North Pacific. To exploit this trade, the French needed to understand what lay between north-western America and China.[1] At this stage the British were still reeling from the loss of their American colonies in 1783. This was the best moment for the French to seize the initiative.

La Pérouse – a grand expedition disappears

Jean-François Galaup, better known as the Compte de la Pérouse, was instructed to undertake a long and ambitious expedition – first to New Holland, to examine the western, southern and northern coastlines left uncharted by Cook, thence to New Zealand, followed by the charting of the north-west American coastline (and the mandatory search for the elusive North-West Passage), the Aleutians, Japan and Kamchatka, part of the Siberian coastline. It was an exhaustive list, and enough for two or three separate expeditions.

La Pérouse was a well-respected, able and popular naval officer who'd been judiciously selected to avoid a repeat of Kerguelen's disgrace.

During the American War of Independence he'd destroyed the British forts deep in Hudson's Bay. Although stung by their defeat, his British enemies respected him highly for the humane manner in which he dispatched the remote, icebound forts. He ensured passage back to England for captured settlers. More importantly, he also left arms, ammunition and supplies for those who'd fled into the forests when the French had arrived or who were otherwise out trapping, and who would've had no refuge from the approaching winter following the forts' destruction.

Back in France he was generally well regarded by both commoner and aristocrat. Despite his wealthy background he'd married the daughter of a government clerk, flouting rigid social conventions. Nevertheless, he'd also managed to charm his way back into the confidence of the aristocracy and the king. This was important. The French Navy of the day was deeply influenced by the aristocracy, more so than the British Navy. Officers of noble birth were even distinguished by their red breeches, with their more common colleagues wearing blue breeches.

Positions on La Pérouse's new expedition were keenly sought. One disappointed applicant was a young cadet by the name of Napoléon Bonaparte. King Louis XVI, himself a keen cartographer, took a strong interest in the expedition. He urged La Pérouse to begin the exploration of Australia along its southern coastline, starting at Van Diemen's Land, rather than running westwards along the northern coastline.

La Pérouse delayed his visit to Australia following his arrival in the Pacific in 1786, exploring the North Pacific instead. During a stay in Kamchatka he sent dispatches back to Paris and also received fresh instructions to proceed to New Holland to ascertain the British intentions there. Despite the loss of the expedition's two longboats and the massacre of his friend and second-in-command, De Langle, and eleven others at Samoa, La Pérouse's humanity prevailed, forbidding reprisals of any sort. The expedition then headed for Botany Bay.

They arrived there on 26 January 1788, just in time to observe the First Fleet of British colonists moving north from Botany Bay to establish itself permanently at Port Jackson. Leaving dispatches with Captain Phillip, La Pérouse built his new longboats and resumed his voyage a month later. With his two ships and a rested crew, he headed north and was never heard from again.

Once more, a well-conceived French plan failed to pass on any knowledge gained and disappeared without a trace, along with all of its extensive natural history collections. Despite the narrative and the charts published from his various dispatches, this was a drastic loss for France.

La Pérouse's contribution to the exploration of Australia's coastlines wasn't to be in the knowledge he added directly but in the contributions of those sent to look for him.

CHAPTER THIRTEEN

*D'*ENTRECASTEAUX'S RESCUE MISSION

'Is there any news of La Pérouse?' was said to have been one of Louis XVI's last questions on the eve of his execution in 1793. Given his popularity at home, La Pérouse's fate continued to obsess the imagination of France, notwithstanding that the French Revolution and a change of government had occurred since his disappearance. Two years previously Louis XVI had sent Rear Admiral Bruny d'Entrecasteaux with two ships, *Récherche* and *Espérance* – crewed by men with loyalties already strongly divided between king and republic – to find and rescue the lost expedition.

Fleurieu drafted d'Entrecasteaux's instructions; he'd also given La Pérouse his. These followed La Pérouse's last known intentions as passed on to the British at Botany Bay. As had been much the case with La Pérouse, d'Entrecasteaux was charged with exploring the remaining coastlines of Australia left untouched by the British, starting with the south coast, then on to New Caledonia, the Solomons and New Guinea, and back to Australia's northern and western coastlines. While at Cape Town on the voyage out, d'Entrecasteaux changed this itinerary, deferring exploration of the southern coastline. Instead they would make a direct passage to Van Diemen's Land to restock with wood and water before making a direct cruise to the Admiralty Islands, where he'd heard of reports of traces of French survivors. He sailed in 1791.

Hunter's wild goose chase

The clues in the Admiralties had reputedly been gathered by Captain John Hunter, returning to England on *Waakzamheyd* following the wreck of the *Sirius* on Norfolk Island in 1790. Hunter was alleged to have told two French captains in Batavia that while passing through the Admiralties he'd seen natives wearing pieces of red and blue clothing, in some of which he recognised the French naval uniform. Also they were making shaving signals – all so uncustomary as to suggest the influence of the French. Contrary winds had apparently prevented closer scrutiny. D'Entrecasteaux was unable to speak directly to Hunter, who later denied making the statements.[1] The difficulty with these reports was that La Pérouse never mentioned an intent to go anywhere near the Admiralties. Hunter's *Historical Journal*, published in 1793, did refer to an Admiralty Islander making shaving signals, leading Hunter to consider whether Europeans had been that way before, and that it might have been La Pérouse. He mentioned nothing about French uniforms. D'Entrecasteaux ultimately gave benefit to the doubt and relied on signed testimonials from French captains received at Cape Town and forwarded to him, and he changed his plans accordingly.

It's possible that had he not changed his plans, d'Entrecasteaux might have achieved his mission. Although d'Entrecasteaux couldn't possibly have realised it at the time, his decision delayed his passage past the final resting place of La Pérouse's ships, in the Santa Cruz Islands, for at least another year. It seems that at some stage, survivors of La Pérouse's ships departed that locality in a vessel built from wreckage and local timber. As it isn't known exactly when they departed, and as they disappeared without trace, one wonders how far they'd got and how often they'd stared sadly as sails in the distance disappeared over the horizon.

D'Entrecasteaux spent some time in Van Diemen's Land in April 1792 in preparation for his deviation to the Admiralties. While there, the expedition discovered among a myriad of islands and bays the secure anchorage of D'Entrecasteaux Channel, which had escaped the notice of Tasman, Marion, Furneaux, Cook and Bligh. This and the surrounding bays and inlets were beautifully charted by his hydrographer, Beautemps-Beaupré, whose work has been regarded since as more accurate than those of any of his contemporary British hydrographers. The charts themselves would become the object of international intrigue following their capture by the British after the conclusion of the expedition.

Van Diemen's Land was now becoming an object of great interest to the French. Marion's 1772 vision of resupplying at Van Diemen's Land had been realised by the British in 1773, on Cook's second voyage, when Tobias Furneaux, in command of *Adventure*, discovered excellent shelter, wood and water at Adventure Bay on the south-east coast of Bruny Island. Since then Adventure Bay had been visited by Cook, on his third voyage in 1777, and by Bligh in the *Bounty* in August 1788 en route to Tahiti. It was clearly of strategic value for any intended voyage to the Pacific that wished to avoid the gales of Cape Horn or difficulties with the Spanish, Portuguese or Dutch governments and settlements by using those or alternative routes. As the southern trade in whale oil and seal fur became more attractive it would have further allure. Beautemps-Beaupré's accurate and intricate charts would have been of great value to any nation.

Leaving Van Diemen's Land, d'Entrecasteaux headed to New Caledonia in May 1792, charting the west coast of New Caledonia for the first time, then through the Solomons, New Ireland, New Guinea and finally the Admiralties by July of that year, before returning to the Australian west coast in December 1792 having found no trace of La Pérouse. Driven past King George Sound – discovered the previous year by George Vancouver in the first stages of his marathon voyage – d'Entrecasteaux followed the south-western Australian coastline to Esperance Bay and the Recherche Archipelago (near the current town of Esperance), named after one of his ships. This was carefully charted for the first time. The expedition continued to the apex of the Great Australian Bight near, but not seeing, the islands of St Peter and St Francis. No European had followed this course since Pieter Nuyts in 1627. There, like Nuyts before him and others since, fickle winds and lack of water forced a retreat in January to the haven of D'Entrecasteaux Channel in Van Diemen's Land.

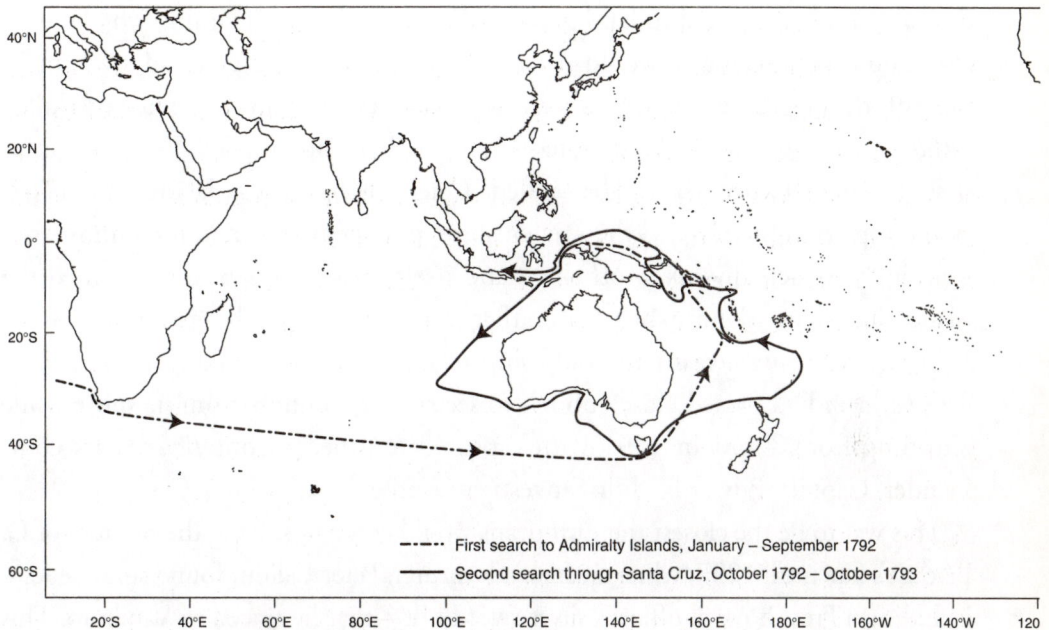

D'Entrecasteaux's two passages around Australia

After five weeks in D'Entrecasteaux Channel and another week at Adventure Bay, d'Entrecasteaux set forth again on 27 February 1793, this time to seek the remnants of La Pérouse in the Santa Cruz Islands, the last reported destination of that ill-fated expedition.

Recherche Island, another near miss for the French

In May 1793 d'Entrecasteaux had his near miss. His two ships passed within 70 kilometres of the final resting place of La Pérouse's ships, *Astrolabe* and *Boussole*, just off a small uncharted island (now Vanikoro) in the Santa Cruz Group, where survivors may still have been alive.

Sadly they saw no reason to stop there. Presumably they reasoned that as Santa Cruz Island, the largest of the group, had been the intended destination of the lost expedition, this is where most information would be found. The islands were also generally renowned for their warlike receptions – as d'Entrecasteaux indeed

101

discovered when he tried to land at Santa Cruz – perhaps explaining why landing at the more remote island wouldn't have been a priority. Carteret had made mention of the aggressive nature of the locals in 1767; they speared and killed *Swallow*'s sailing master. He'd noted Ourry's Island but missed Vanikoro as he passed Santa Cruz, which he called Queen Charlotte's Island, in 1767. Beautemps-Beaupré charted the French ships' passage past Vanikoro; a number of bearings were actually taken off the island itself. Beautemps-Beaupré named the island Recherche Island, which was rather ironic given how things turned out.

D'Entrecasteaux's wasn't the only ship to pass close to Vanikoro at that time. The *Pandora* had passed close enough to see smoke columns from its shore while returning from Tahiti in 1791 with captured mutineers from *Bounty*. Its commander, Captain Edwards, didn't investigate, either.

This was to be the closest the Europeans would come to solving the mystery of La Pérouse's fate until 1827, when an Irish adventurer, Peter Dillon, found some relics – including a French naval officer's silver sword hilt – that he traced to Vanikoro. This led to the discovery of some remains of one of La Pérouse's vessels neighbouring the reef she and her consort had struck in a cyclone almost forty years before. Even today, not much more is known of their fate. It appears that survivors of the wreckage built a vessel and left the island, but no-one is sure of time of departure or destination. They left no trace other than the relics of shipwreck, cut stumps and the remains of a stockade. We can only speculate as to whether they'd stared sadly as passing ships ignored their signal fires and left them to their isolation.

For d'Entrecasteaux's hapless mission, sickness was now taking its toll. Only two months after passing Vanikoro, d'Entrecasteaux himself died of scurvy, shortly after his second-in-command, Huon de Kermadec. The two ships passed through and charted the islands off the north-eastern coastline of New Guinea, headed around the northern side of New Britain and then south to Java. The cataclysmic news of the French Revolution, the execution of Louis XVI and France's war with most of Europe awaited them when they arrived at the Dutch settlement at Surabaya in late 1793.

Beautemps-Beaupré's chart including Recherche Island (now Vanikoro), lower right, with the track of the expedition also marked

The collapse of the expedition

Here was an exhausted expedition, with the loyalties of its members fiercely divided between royalist officers on the one hand – who'd been commanded by their now dead leaders, d'Entrecasteaux and Kermadec, under the commission of a now dead king – and on the other the republican body – many impressionable junior officers, crew members and 'savants', whose heroes now wielded power in Paris. The savants, twelve in all – comprised of artists, astronomers and hydrographers – were particularly influential men in the new French republic. The expedition was now financially orphaned, given the demise of Bourbon France, and the fact that its republican successors were now enemies of the expedition's colonial Dutch hosts and of questionable creditworthiness. Suspected republicans on the expedition were interned. The Dutch eventually seized the expedition's two ships, *Récherche* and *Espérance*, from d'Auribeau, the surviving commander, to cover port and other costs of feeding and deporting its members. D'Auribeau died shortly after, and much of the expedition's natural and hydrographic product as well as official logbooks, charts and journals fell to the safekeeping of Lieutenant Rossel, former first lieutenant of *Récherche*.

As a royalist, Rossel – like many others – would have faced the guillotine upon his return to France. Fortune smiled on him. With Beautemps-Beaupré's charts and the botanical and geological specimens from the expedition firmly in possession, Rossel, on the Dutch ship *Houghly*, was part of a convoy sailing back to France when he was captured by the British. The Netherlands had by this time fallen to the French, so they were now an enemy of the British. Rossel and the collections were moved to *Sceptre*, the British frigate now escorting the convoy to Britain. This was lucky as while still in convoy *Houghly* sank in a gale shortly after Rossel was transferred. For the British his capture was a coup, but wasn't capitalised upon. The charts remained unpublished for the time being, and the botanical and other specimens were returned to Paris through the diplomatic auspices of Sir Joseph Banks. The expedition's surviving republican naturalist, Labillardière, returned from Batavia to Paris in 1796 to arrange for publication of an unofficial narrative of the voyage. It was he who negotiated with Banks for the return of the natural collections. The charts of the voyage and the official narrative were published in 1808 following Rossel's return to France. They are today a rare item.

Once again, the product of a well-planned French expedition had been lost and its commanders were dead. The experience of its officers and men had been scattered as a result of their clash of political beliefs and the chaos of events that had ensued when the strife of the French Revolution had reached from the streets of Paris right to the decks of the expedition's ships. Anyone of influence who wasn't dead from the voyage or after imprisonment in the filth of a Dutch colonial jail had either died of dysentery in the Dutch convoys to Europe or gone into exile abroad, with few returning home.

CHAPTER FOURTEEN

BAUDIN, THE FIRST BIG FRENCH SUCCESS

The first major achievement of the French to have lasting effect was Nicolas Baudin's expedition from 1800 to 1803. Baudin is one of those characters to whom one suspects history would have been kinder had he survived this expedition to defend his character.

Although aged forty-six when the expedition commenced, and without any hydrographic training at all, Baudin had had many years' sailing experience with scientific voyages, initially in the service of the Austrians. His orders from the French Government were to survey the uncharted southern and northern coastlines of Australia, including the uncharted Torres and Bass straits, as well as southern New Guinea. He was also to look for a north–south strait dividing Australia. The French had the same suspicions as the British of a north–south strait dividing the mainland, and the same desire to discover new country.

Baudin's expedition would be one of the major reasons for the formation of Flinders' rival British expedition on *Investigator*. That Baudin was able to complete his tasks with his meagre background in survey work was an achievement. That his efforts led to the creation of a French equivalent of 'Cook's school' of navigators was far more noteworthy.

Early days

Baudin's past is interesting. In the service of the Austrians he'd survived shipwreck and cyclone and had developed a reputation for successful delivery of plant specimens from abroad to the Austrian royal family. During the French Revolution he'd offered his services to the French Navy as a vessel captain. The navy didn't wish to engage him at this rank so he took his services back to the Austrians, the enemies of republican France – but not of a refugee sailor with royalist sympathies – for most of this period.

He first came to prominence with the French by successfully proposing to sail to Trinidad in 1796 to retrieve a cargo of natural history specimens for the Museum of Natural History in Paris. His history of service to the Austrians didn't prejudice this, given his promises to effectively deliver to the French a cargo collected by him for their enemies. It appears to have been Baudin's responsibility to arrange all transport. If nothing else, this trip demonstrates the man's tenacity. Because he had little financial credit, he borrowed a boat for the first leg of the trip. It only just survived a major storm to reach Tenerife so he replaced it there with another. He did some botanising in Tenerife, then when he reached Trinidad, he found it'd fallen to the British, then enemies of France. The British refused to recognise his safe conduct passport or release the cargo. Undaunted, he exchanged his ship for yet another, bigger one and botanised in surrounding islands and in Puerto Rico, compiling a substitute collection of specimens, including a palm tree, before returning to France in 1798.

His luck seemed to mirror that of Flinders on *Reliance* over the same period; he arrived in France just in time to have a selection of his 3500 exotic plant specimens join a procession of military prizes of the young General Napoléon Bonaparte along the Champ de Mars in Paris. With this profile and the money from sales of plants to collectors, including the Museum of Natural History, Baudin was suddenly well placed to promote a further expedition. He was shortly afterwards promoted to *capitaine de vaisseau* and now had powerful allies in the Museum of Natural History, particularly its influential director, Jussieu.

Baudin's 'bold dash'

With a style that Flinders would replicate with his 'bold dash forward' two years later, in July 1798 Baudin submitted a proposal to the Ministry of Marine, with museum director Jussieu's backing, for a new voyage around the world with three ships. He claimed in his plans that among the plants of New Holland was one that alone would justify the expedition – probably flax, required for making rope – but the explanatory memorandum setting out what this was has since been lost. Baudin further claimed that the voyage would resolve speculation on whether New Holland was one island or more. Flinders would also argue this in his later proposal to Banks. Baudin's idea was shelved until 1800 when, following Napoléon's coup d'etat and effective seizure of power, Baudin wrote directly to Napoléon on the subject.

This was an audacious move. Napoléon passed the request on to the minister of Marine. Baudin widened his audience and enlarged upon his plan, lobbying the Institut National, the major support base for such projects. The institut referred the proposal to a commission consisting of Fleurieu, Bougainville, Jussieu and four others, which supported a revised version. The revised plan recommended by the commission pertained only to New Holland and the southern coastline of New Guinea. As the British penal colony at Port Jackson was by this time firmly established, the French were interested only in unexplored and unsettled coastlines of New Holland, and in any harbours and rivers that might enable penetration of the unsettled inland. The commission believed a coastal survey of both the Torres and Bass straits to be sufficient justification for the expedition. However, in the interim, the British had published charts of Bass Strait, meaning that the British claims to New South Wales could arguably not apply to the separate and unsettled Van Diemen's Land or to any part of New South Wales, as claimed by the British, lying west of and separated by a north–south strait. This may have also heightened French interest in Van Diemen's Land.

Napoléon met a delegation, including Baudin, on 26 March 1800. On 3 April an outline of the expedition was presented to the First Consul, who signed it only five days later.

Baudin had been able to combine his lucky break in 1798 with political circumstances to create a further opportunity for himself. On 29 April 1800 the minister of Marine, Forfait, recorded that the First Consul was well disposed to order a voyage of exploration of the south-west coast of New Holland, where Europeans hadn't yet visited. Baudin's 'bold dash forward' had paid off, and a passport for the voyage was requested from the British Government in the following June.

The French initiative would provide the incentive for the British to approve the similar and in no way coincidental proposal made by Flinders for the *Investigator* expedition only some seven months later, particularly once passports of safe passage for Baudin were duly sought from the British.

Speculation has abounded since as to why Napoléon granted his approval in only five days. Did this reflect French territorial aspirations? Did a young and successful Napoléon desire to acquire and settle parts of Australia? Probably not, although the information gathered by the expedition would be useful in the future. At that stage Napoléon was in only the first few months of his newly won position as First Consul. The Institut National, the focus of French scientific thinking, was a powerful and well-respected French institution, the members of which had by and large avoided the terrors of the revolutionary guillotine. It was the closest thing the French had to Joseph Banks and the Royal Society. Napoléon's approval of a proposal such as Baudin's served to pacify this still influential faction as the young leader consolidated his power. Quite simply, the reasons for the approval may have gone no further than this. Napoléon's recent Egyptian campaign had failed following Nelson's annihilation of his fleet at the Battle of the Nile. He would've been unlikely to seize upon another such naval venture, certainly not one in a more distant land like Australia that was even more reliant on naval support. Moreover, his attention was focused on domination of the European stage, where military rather than naval strength could prevail. One must assume that this voyage was an early reconnoitre, motivated more by curiosity. Its dominant purpose was scientific rather than strategic.

Baudin's scientific and political burdens

The scientific goals of the expedition provided Baudin not only with purpose but also with his greatest burden, in the form of twenty-two civilian scientists and artists. This was six more than Baudin had originally estimated, and more than three times what he ultimately asked for. Although three of them, assistant botanist Riedlé and zoologists Maugé and Levillain, were Baudin's trusted ship-mates from past voyages, the others knew little of Baudin. One of them – François Péron – would become the chief assassin of Baudin's character and ultimately assume most of the credit for the expedition.

Péron

As the son of a village tailor, Péron came from humble beginnings. Despite partial loss of vision in one eye from childhood smallpox, he excelled at the local school. With his teacher's support he'd been in training for the priesthood when the fervour of the Revolutionary Wars finally reached his small village in 1793. He volunteered for service with the local musketeer regiment. Under siege at Landau for eight months against the Prussians, eighteen-year-old Péron stood out as courageous and intelligent, and was quickly promoted to sergeant. The appalling conditions caused by the siege, fatigue, non-existent hygiene and the primordial behaviour this generated in his comrades towards prisoners, civilians and others left their marks. He lost all sight in his weak eye as a result of the unsanitary conditions. He was subsequently wounded, taken prisoner, released and invalided out of the army, much against his wishes, due to his partial blindness.

With no resources of his own, he returned to his home village. But this wasn't to be the last the world heard of him. Having now given away ideas of the priesthood, he looked for another calling and the means by which to follow it. Finally he won support from a wealthy local notary to study medicine in Paris in 1796.

This was a big break for Péron and he absorbed himself in the study of all facets of natural science and history as well as medicine. Typically, he developed an excellent rapport with his professors, including botanist and museum director Jussieu, zoologist de Lacépède, and physiologist Bichat. Péron was also influenced by zoologists Lamarck and Cuvier. Lamarck was the first to propose the idea of 'spontaneous generation' of life forms and of modification of species, half a century before Darwin and Wallace developed their theory. Cuvier had founded the disciplines of comparative anatomy and palaeontology that would ultimately place man in the animal kingdom rather than as a creature purely of God's creation.

By the time of Baudin's 'bold dash' in 1798 Péron had been culturing these contacts and his own capabilities for two years. The French were at this time developing a separate field of study – anthropology, the study of man. The Society of the Observers of Man was founded in 1799 for this purpose.

Baudin's expedition was to research, among other things, the physiology and customs of local inhabitants, and specific scientists were appointed to this task. This was something utterly new. Cuvier was one of those responsible for instructing expedition members on this subject. He was also the man who'd introduced the study of comparative anatomy. Péron, applying too late for an existing position on the expedition, persuaded the organisers to accept him for an additional position: trainee zoologist charged with comparative anatomy. As no other zoologist on the expedition had training in medicine, he seemed the right man for the position. His well-groomed contacts with Jussieu and Cuvier helped tip the balance in his favour. It was Jussieu who'd originally supported Baudin's 'bold dash'. Péron's innate networking capabilities facilitated his joining an impatient Baudin on *Géographe* on the eve of departure.

Péron was a man of amazing drive and ambition, perhaps sourced in his humble background. He also had a great ability to charm and a desire to ingratiate himself to others at all levels of power to achieve his own ends, as he'd demonstrated so far in order to secure a place on the vessels, and as he would demonstrate again in the future, both to Baudin's and Flinders' disadvantage. Coupled with his undoubted intelligence, these qualities allowed him to make the most of any opportunities that presented themselves. One such opportunity would ultimately be the publication of the expedition's scientific exploits. Péron would become obsessed by a belief that this was necessary for his own advancement, much to Baudin's disadvantage.

The ambitious Péron wouldn't be Baudin's only problem.

Baudin's young gentlemen

The most impressionable element of the navy's officer corps, its midshipmen, were another curse for Baudin. For the British and the French alike, midshipmen were usually young and untrained; they were expected to learn from the experience the cruise offered and from that to further their careers. They formed the pupils in 'Cook's school'. Vancouver, Clerke, Bligh and others had learnt from Cook as midshipmen and master's mates, as had Matthew Flinders from Bligh, and Samuel Flinders and John Franklin from Matthew Flinders.

A position as midshipman on an expedition of importance was keenly sought by the rich and well-connected for their young, aspiring, but often over-indulged and totally inexperienced sons. It was a means of learning from, and gaining the patronage and support of, those of influence in the navy, which was so important for promotion and a successful career. By the same token, the presence of so many immature, restless and influential characters on a ship needed to be closely super-vised by experienced officers. If well handled and properly educated, they were a navy's lifeblood. Otherwise, they could be dangerous. Bligh blamed the *Bounty* mutiny in part on impressionable, young and inexperienced midshipmen and master's mates exposed to the delights of Tahiti and to devious seamen, in tandem with the lack of any competent senior officers – himself excepted – to supervise them. Vancouver's demise was largely due to his inability to properly deal with a young and influential midshipman, the future Baron Camelford.

Baudin had asked that he have no midshipmen. He was given fifteen. One of these was Bougainville's son, Hyacinthe, an influential source of dissension. The burden of fifteen young midshipmen coupled with the presence of twenty-two 'enlightened' civilian scientific staff used to questioning everything and used to rational behaviour and argument in response, rather than the autocracy of naval discipline, was to cause endless disharmony.

Mauritius and malcontent

Baudin was given command of two corvettes for the voyage, the *Géographe* and the *Naturaliste*. They were lavishly equipped. However, they'd been supplied with provisions for only nine months, requiring Baudin to call in at either the Cape or Mauritius for further supplies to support the long voyages off the Australian coastlines. Baudin was also directed to acquire a third vessel en route, of shallow draught for close inshore work. Timing was fairly tight. If they lost much time on the passage out they'd miss the favourable winds required to chart the south coast of Australia, as his instructions required.

Unfortunately this is exactly what happened. Baudin was sailing at a time of war between Britain and France. The British captured the Cape in 1795 from the Dutch – and weren't to return it until 1802 under the Treaty of Amiens – so Baudin decided to resupply at French-controlled Mauritius. This was an unfortunate choice. Mauritius was an island under British blockade. Supplies came largely from its own meagre resources and from the plunder of British merchantmen by French privateers based there. The local administration not only wouldn't assist in resupplying Baudin, but a concerted effort was made by the locals to entice his crews to desert and crew their privateers' fleet. Further, both of Baudin's ships would have made a useful addition to the colony's sparse naval fleet, should the expedition have been unable to continue. Not only could Baudin not obtain his shallow-draught vessel or even adequate supplies, he also had great difficulty holding his existing expedition together.

At Mauritius Baudin lost experienced members of both his officer corps and crew, which would have caused him concern, as well as individuals from his scientific staff, which would have caused him less. The departing officers included one of his most experienced, Lieutenant Pierre-Guillaume Gicquel – who'd sailed with d'Entrecasteaux – and Lieutenant Bertrand Bonnié. Generally, the official reason given was ill health. Some of the real reasons for the departures may have been markedly different, and influenced by the manipulative Péron.

There had already been seven biologists – three botanists and four zoologists – on board when Péron arrived. His unexpected presence, through the efforts of Cuvier and Jussieu, would have caused a degree of professional jealousy. This may have been antagonised during the long journey to Mauritius, particularly when the ships were becalmed for some weeks off the African coast. Whereas some of the scientific staff may have been prepared to do nothing towards their appointed tasks until their arrival at New Holland, Péron exhibited boundless enthusiasm

and energy. Unaffected by seasickness, he was constantly about the ship conducting all types of measurement. He developed a unique interest in studying the sea
itself, probably the first serious attempt at oceanography since John Forster's measurements on Cook's second voyage. Péron made six-hourly recordings of temperature and salinity at different depths throughout the whole voyage, often to
Baudin's critical amusement. These he compared with measurements of surface air
pressure, temperature and humidity. He also developed a totally new interest, the
study of marine invertebrates. His mentor, Cuvier, had pioneered the study of
land invertebrates, and Péron pursued its maritime counterpart with characteristic vigour.

Péron may well have appeared as an upstart to the scientific staff and the senior officers, particularly those who felt threatened by his confidence. He would
have come across as a wounded hero of the Revolutionary Wars to the young and
impressionable midshipmen and other junior officers. Baudin couldn't match this.
Rather than serving the French Navy and the Revolution during that time –
despite the fact that he'd offered to do so – his occupation would have appeared
more selfish and entrepreneurial as a specimen collector for their one-time enemies, the Austrians.

Both the expedition's artists left at Mauritius. Baudin replaced them with two
artists he'd already hired, although entered on the ship's lists as 4th class gunner's
mates. They were Charles-Alexandre Lesueur and Nicolas Martin Petit. Péron
became firm friends with Lesueur, who painted many impressions of Péron's specimens. His and Petit's work, however, while of great artistic value, was unfortunately relatively useless to botanists and anatomists, as neither had any knowledge
of the scientific detail or presentation required of drawings from a voyage such as
this. For example, a now celebrated, beautifully coloured and executed painting of
a fish by Lesueur bears the title *Poisson Inconnu* – 'unknown fish'. The point is particularly clear when these are compared with the fine works of Ferdinand Bauer
from Flinders' *Investigator*.

Although Baudin had difficulties enough keeping his expedition together and
also obtaining the supplies he needed at Mauritius, he didn't help his position with
the locals by suggesting to the governor that his expedition had political rather
than scientific aims, in order to encourage assistance. The governor didn't believe

him. For Baudin to suggest this risked invalidation of his passport from the British Government. It only increased the governor's suspicion of his motives and justified in his mind the reports critical of Baudin that he'd sent back to Paris. It has also fuelled massive academic speculation since as to whether Baudin's real orders actually were political, namely to assess Australia for French occupation, rather than scientific. In any event, the tangible result of Baudin's ploy was to cause crucial delay.

The Australian west coast

By the time Baudin's two vessels reached Australia in late May 1801, he was short of good water and fresh food, and more than two months behind his timetable. Although he was scheduled to first chart Van Diemen's Land then return to the west coast, he decided to focus initially on the south-west of Western Australia, at Geographe Bay, where he could also replenish supplies of water and wood. The wreck of the *Géographe*'s longboat, the drowning of a seaman, Vasse, at Geographe Bay, and the onset of a series of severe gales led to the hurried departure of the expedition from Geographe Bay and subsequent separation of the two vessels as they attempted to chart the west Australian coastline as far north as North West Cape. *Naturaliste* lost a series of anchors as it scampered out of danger from these gales in Geographe Bay, leaving it with only one spare large anchor stored deep in its hold, accessible only in fair weather at a safe anchorage. Separation was to be a constant state for these two vessels. As a precaution, Baudin and *Naturaliste*'s commander and Baudin's second-in-command, Emanuel Hamelin, pre-arranged to rendezvous, first at Rottnest Island and then Shark Bay further up the coast.

At Rottnest Baudin failed to see *Naturaliste* waiting for him while her precious remaining bower anchor was hoisted up from the hold. Baudin sailed on to Shark Bay and therefore missed participating in the rediscovery by Hamelin's officers of the Swan River, first found, and last seen – by European eyes – by the Dutch explorer de Vlamingh over a hundred years earlier. Hamelin didn't catch Baudin further up the coast, either; the vessels finally met in Timor. Baudin had left Shark Bay by the time Hamelin reached it.

Not knowing if Baudin was behind or in front of him, Hamelin stayed some weeks at Shark Bay as well and luckily managed to explore almost all the areas not visited by Baudin in the weeks prior. His men discovered de Vlamingh's pewter plate in the sand on Dirk Hartog Island. This plate had been left by de Vlamingh in 1697, in turn replacing one his men had found there that had been left by Dirk Hartog in 1616. Hamelin ordered his men to return de Vlamingh's plate, nailing it up on a new post. It was removed to Paris seventeen years later by one of Hamelin's junior officers, Louis de Freycinet, on a later expedition up that coast. Here also another anchor lost a fluke on a clean sandy bottom, leaving Hamelin with only one serviceable large anchor and causing words of concern to be muttered. Twenty years later Phillip Parker King would mutter similar words on the quality of naval stores and foundries for the same reason, on the same coastline, when also left with only one anchor and forced to terminate his survey.

Baudin, meanwhile, had headed north to North West Cape, the Dampier Archipelago and the Bonaparte Archipelago. Here, in search of water, they visited Depuch Island, named after the expedition's mineralogist, Louis Depuch. Despite discovering the island and a water source, his men missed its beautiful Aboriginal paintings; the British discovered them thirty-nine years later. Baudin's plans to head for the Gulf of Carpentaria were thwarted by the changing seasons, shortages of wood and water, and the first signs of scurvy among ten of his crew. On 19 August 1801 he turned for Timor.

Baudin's dissatisfaction with his junior officers and some of the scientists was showing by this stage. In his sea journal and on his preliminary charts he'd even named landforms after their failures, with Anse des Maladroits reflecting commander Le Bas's loss of *Géographe*'s longboat (near what is now Wonnerup Inlet in Geographe Bay), and Cap des Mécontents (now Cape Naturaliste) commemorating a feature that Sub-Lieutenant Picquet had failed to reconnoitre and take bearings on, despite staying out one night.

Scientists got lost ashore, missed deadlines and refused to follow orders so frequently that Baudin wouldn't allow them ashore unless accompanied by his chosen officers. With so many scientists this must have been a major headache for him. The shallow waters of the northern coastline didn't lend themselves to close inshore examination, particularly without the benefit of the longboat lost at Geographe Bay. As he retired to Timor, Baudin knew he'd need to return for a second visit to

Louis Antoine de Bougainville, the first to investigate the Australian east coast, and patron to voyages of many French navigators.

(Francois-Seraphin Delpech, National Library of Australia)

Jean-François Galaup, Compte de la Pérouse. La Pérouse disappeared with his entire expedition, the French ships *Boussole* and *Astrolabe*, in 1788.

Rear Admiral Antoine Bruny d'Entrecasteaux was charged with the task of rescuing La Pérouse by King Louis XVI in 1791. He died of scurvy in 1793.

Captain Nicolas Baudin was commander of the French expedition of 1800. Despite his efforts in securing the success of the expedition, his name was not mentioned in its published records or charts.

(François Bonneville, National Library of Australia)

François Péron, naturalist on the Baudin expedition, and the chief assassin of Baudin's character.

(Ambroise Tardieu, National Library of Australia)

An Aboriginal woman and child from New South Wales – the Baudin expedition was one of the first to interact closely with Australia's Aboriginal people.

(Barthelemy Roger, National Library of Australia)

Sir Joseph Banks KB, president of the Royal Society, essential patron of the British navigators and ex officio minister of sciences and the colony of New South Wales.

(William Daniell, National Library of Australia)

Captain James Cook, whose efforts laid the foundations for the successes of the British navigators.

(James Basire, National Library of Australia)

Above: Rear Admiral William Bligh, the great survivor. The only British navigator to achieve what could be called 'comfortable retirement'.

(Alexander Huey, National Library of Australia)

Right: Captain Matthew Flinders – a brilliant, ambitious but possibly immature navigator, he would be the first to promote the name 'Australia'. His charts were the most accurate for many years.

(National Library of Australia)

Above: Breadfruit. Transplanting this fruit from Tahiti to Jamaica was crucial to both Joseph Banks and William Bligh.

(Possibly by Sydney Parkinson, National Library of Australia)

Left: The enterprising and energetic George Bass, discoverer of Bass Strait and Western Port, who is thought to have died either in the southern oceans or the Spanish silver mines of South America.

The wrecks of *Cato* (left) and Flinders' *Porpoise* (centre right) at Wreck Reef in 1802.

(William Westall, National Library of Australia)

Above: The tiny *Lady Nelson*, the first ship to sail into Port Phillip and along the Victorian coastline. After over twenty-five years of service, she and her crew were destroyed by pirates north of Timor in 1825.

Left: General Charles Mathieu Isadore Decaen, governor of Mauritius. A favourite of Napoléon, Decaen was responsible for Flinders' detention on Mauritius for six-and-a-half years, after their one, disastrous, interview.

complete the survey. Nevertheless, Baudin's work along this coastline would remain the most detailed survey of this area until Phillip Parker King's surveys some twenty years later. Flinders would never view it close up, having greater concerns – as we'll see – than inshore hydrography when he sailed this coastline in 1803.

Timor and its appalling legacy

Baudin's extended visit to Timor while a replacement longboat was under construction saw the build-up of further tensions. It introduced dysentery and other tropical diseases that wouldn't only take their own toll but also hasten the onset of scurvy in his ships.

At Timor Baudin was threatened with a sword by Sub-Lieutenant Picquet, whom he'd removed from a watch while still off the Australian coast. Baudin now took the opportunity to remove him from the expedition. This generated much disquiet among Péron and officers who openly took Picquet's side, including Baudin's second-in-command, Le Bas. Picquet was well connected, through the head of the Ports Office in France, and Baudin needed to be careful. But Le Bas also accused Baudin of spending on himself money intended for the expedition's officers. This together with the insubordination in openly supporting Picquet meant Le Bas had to go, too. A duel that then took place between Le Bas and Ronsard, a more loyal junior officer, caused wounds to Le Bas. These, along with syphilitic symptoms and some other complaints that the ship's doctors, at Baudin's instigation, had discovered, were instrumental in Le Bas agreeing to quit the ship. Le Bas was the third experienced officer to depart the expedition, following Gicquel and Bonnié at Mauritius, leaving most responsibility for ship management on Baudin's shoulders alone.

In a further blow to Baudin, he lost his most prolific botanical collector and one of his closest friends from past voyages, Riedlé. He contracted a fatal fever while botanising ashore at Timor; he'd been trying to make up for the lack of opportunities for shore visits along the north Australian coastline, caused by the shallows and lack of a longboat. He was buried in Timor alongside David Nelson, the Banks-appointed botanist from the *Bounty* who, as we've seen, had survived Bligh's open-boat voyage only to die in Timor shortly after, over ten years previously.

Dysentery would also shortly strike down Baudin's only other close friends, zoologists Maugé and Levillain, leaving him almost totally isolated with his volatile command.

Baudin's character, and in particular, his dry and ironic sense of humour, would've only made things worse for him. For example, at one stage on the voyage to Mauritius one of the scientists, the astronomer Bissy, had asked for a spare compass needle. When these were produced they were found to be rusty and useless. Baudin's comment to Bissy was typical:

> Well what can you expect? All the equipment provided by the Government is of the shoddiest quality. If I had my way I would have had needles of silver instead of steel.[1]

The fact that although not likely to rust, they wouldn't be of any use without the magnetic qualities of iron was the joke, but Bissy took the comment seriously. Bissy left the expedition at Mauritius, and the remark was published back in France shortly after. Baudin's joke was reproduced time and time again as evidence of his lack of understanding of something as simple as magnetism. It totally overlooked the humanity of the man or his dry wit.

Baudin's jokes might have been able to relieve the tensions and rigours of eighteenth-century sailing life. Certainly this would have been the case among Baudin's shipmates from past voyages, Riedlé, Maugé and Levillain, but as they died one by one, these tensions would only have been aggravated.

Van Diemen's Land and the unknown south coast

On 13 November 1801, with the *Géographe*'s new longboat stowed aboard, the two ships sailed in company from Timor to Van Diemen's Land. They did further survey work in the area around what's now the Derwent River and also surveyed the island's east coast. There they sorted out the years of confusion surrounding the location of Frederick Henry's Bay plus the shape of the Tasman Peninsula and its two narrow isthmuses, which had confounded Cook, Bligh and d'Entrecasteaux.

Members of the expedition spent a considerable amount of time with the Tasmanian Aborigines, both on the mainland and at Bruny Island. Petit did a number of sensitive sketches of these people that are now a precious record of that since-exterminated race. After this interlude the expedition proceeded north to chart Maria Island, where René Maugé, Baudin's only surviving shipmate from past voyages, finally succumbed to dysentery from Timor. He was buried on the southern point overlooking Oyster Bay on the western side of the island.

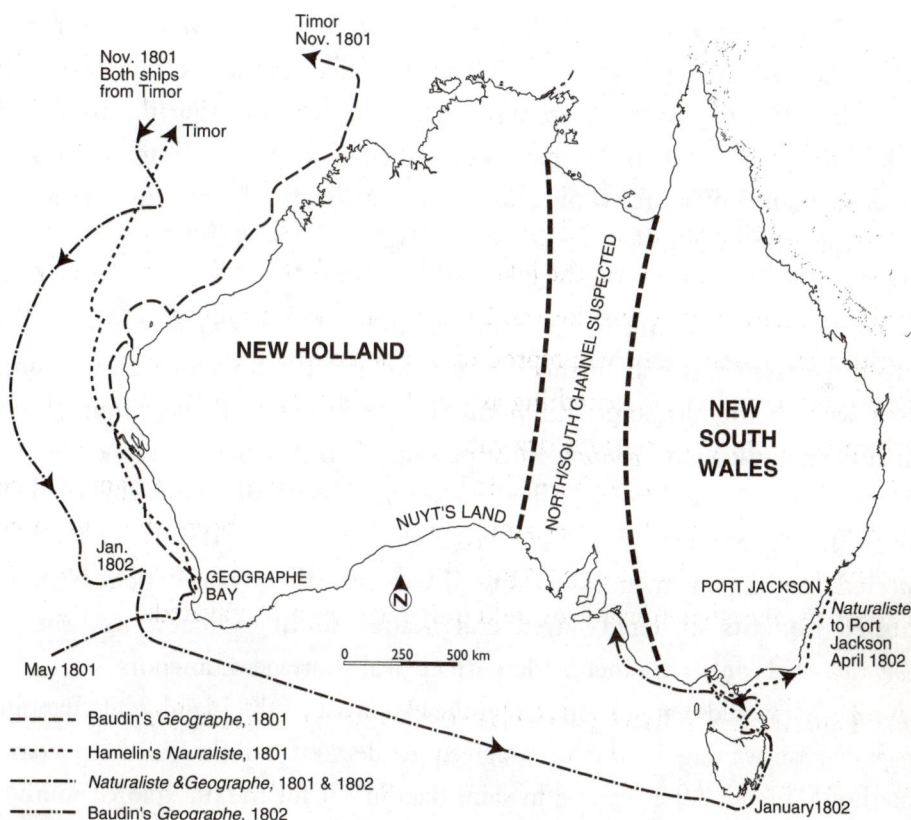

Baudin's expedition to April 1802

More bad news was to follow. On 6 March 1802, north of Maria Island, Baudin dispatched his hydrographer, Charles-Pierre Boullanger, in one of the ship's boats – accompanied with a midshipman and six men – to get a closer view of the coast because Boullanger was 'unfortunately very short sighted and can only take bearings and angles with his nose on the ground'.[2] Although he was given instructions to stay in sight of the ship and be back by nightfall, the dinghy didn't return.

While communicating the loss to *Naturaliste*, the two ships collided and *Naturaliste* broke *Géographe*'s spritsail yard. Three days of searching produced nothing except another separation from the *Naturaliste* after it rapidly headed off in an opposite direction to *Géographe* during the search. The two ships wouldn't meet for another four months. As for the missing dinghy, Boullanger and his crew were picked up on 9 March 1802 by the British brig *Harrington*, the same ship that was later said to have news of Bass's fate in Chile, after three days at sea. They were delivered safely to the *Naturaliste* at a fortuitous meeting the next day in Banks' Strait, the small strait separating mainland north-east Tasmania from the Furneaux Islands offshore, while *Naturaliste* was fruitlessly awaiting yet another rendezvous with *Géographe*.

The *Naturaliste* then continued its search south for the *Géographe*, which Hamelin correctly guessed was still further south searching for Boullanger. Luck simply wasn't with them. Although *Naturaliste* had anchored in Banks Strait, through which Hamelin knew *Géographe* must pass, a gale blew up and snapped both her cables, forcing Hamelin to seek shelter elsewhere and losing both anchors in the process. In the meantime *Géographe* passed through Banks Strait, as Hamelin had expected, heading north, and the ships missed each other in the dark as *Naturaliste* returned from its sheltered haven and headed south to renew its search for *Géographe* with only two anchors left – a critical shortage of anchors.

Naturaliste would search south along the east coast of the island while Baudin in *Géographe* sat waiting at the next agreed rendezvous on the northern coast at Waterhouse Island. After waiting in vain, Baudin left for the unexplored south-east coast of the Australian mainland. *Naturaliste* would finally abandon its search south, and instead head for Port Jackson, surveying Western Port and Wilsons Promontory on the way, but – crucially for Baudin – carrying with it both the expedition's hydrographers, Faure and, although Baudin didn't know this, Boullanger.

This left Baudin in an unenviable position. Without hydrographers and not knowing what had happened to *Naturaliste*, he had only the support of his young

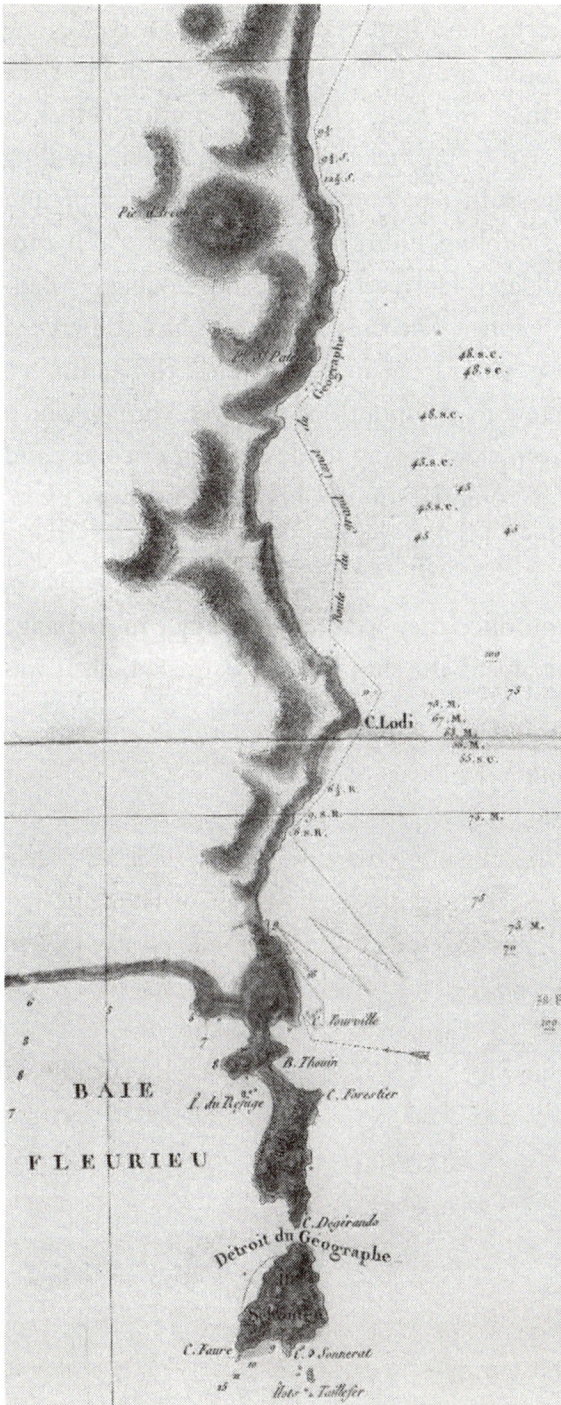

Boullanger's chart of part of the east coast of Van Diemen's Land – including Freycinet Pensinsula and Schouten Island – marking his track in the dinghy

and inexperienced junior officers for the most important part of his survey, along the uncharted south Australian coastline west of Bass Strait to the head of the Great Australian Bight. This was where the hope of a north–south channel or strait lay. Baudin was obviously after this 'big ticket' discovery, a strait dividing Australia, which could alone generate sufficient notoriety for the expedition and the discovery of which, Baudin was gambling, didn't require the services of hydrographers. They could fill in the detail later. However, the missing Boullanger hadn't only deprived Baudin of a hydrographer. Their search for him had also wasted precious time, the value of which was about to be made very clear to Baudin.

By 8 April 1802 *Géographe* was close to the mouth of the river Murray, which they failed to see as it was hidden from view, having made no such discovery and with the prospect and excitement of potential ground-breaking discovery close. Then they sighted a sail in the distance, heading towards them.

The missing *Naturaliste*?

Sadly this hope faded all too soon once they were close enough to exchange identities with the strange ship. This wasn't the ship Baudin wanted at all. It was Flinders' *Investigator*.

CHAPTER FIFTEEN

$\mathcal{F}$LINDERS' BLUNDERS, AND TRAGEDY

Much had happened to Flinders since his successes in London, most of it bad. In fact he'd made a couple of big political errors since his appointment to *Investigator*.

The blunders

Flinders' first blunder wasn't so much what he did but rather how he did it. He got married to Ann Chappelle. But he didn't first tell Banks. Given what Banks had just organised for him, and how much Flinders' future career would depend on Banks, this was incredibly rash. In the navy, marriage was usually something that was cleared with superiors if a large expedition was pending and the marriage was likely to cause distraction. Flinders' secret plan was that Ann accompany him to Australia in *Investigator*. This is evident from correspondence with her just ten days before their wedding, where he suggests that *Investigator* has space enough on board for his wife, 'with love to assist her, to be happy'.[1] It also exhorts her to keep the idea a secret:

There are many reasons for it yet, and I have also a powerful one. I don't know exactly how my great friends might like it.[2]

Banks didn't like it one bit. In fact he only found out about the marriage over a month after it had happened, by reading about it in the local Lincolnshire newspaper.[3]

It seems Ann expected that she could accompany Flinders at the time they married, although Flinders would've known this would cause eyebrows to be raised in the Admiralty. Certainly no wives or family made the voyage to Australia on navy ships without Admiralty permission.

This may well have led to the next blunder. While *Investigator* was sitting at anchor at the Nore prior to departure for Portsmouth, legend has it that the newly appointed First Lord of the Admiralty, Earl St Vincent, decided to pay her young commander a surprise visit.[4] The earl, with whom Flinders didn't have the same support or interest as his predecessor, Earl Spencer, would've been curious about this rising young star, Flinders. Clearly discipline was lax on *Investigator* – no lookout reported the arrival of the admiral's launch; no rope boys were provided to greet him; and, worse still, when entering the cabin unannounced, there was Flinders with his wife sitting on his knee *without her bonnet on.* To the old Earl, this was tantamount to her having taken up residence on the vessel, which would have been forbidden without Admiralty permission. The admiral left, mightily unimpressed. By then Flinders had written to the Admiralty asking permission for Ann to go with him on *Investigator.* While this decision was pending, on 28 May 1801 he sailed *Investigator* from the Nore to Portsmouth – with Ann on board.

A final, major blunder then occurred that decided the matter for Flinders even before the Admiralty could. He ran the *Investigator* aground. This was a serious error. *Investigator* hit a sandbank at a place called The Roar (now Roar Bank), off Hythe. Admittedly, although it had been charted, it wasn't on the chart supplied to Flinders. Being early June it was still daylight, with light winds and, luckily for Flinders, a rising tide, meaning that he got off without damage in a couple of hours. At the time, only young Samuel Flinders was on watch; Flinders was below with his wife, and there was no pilot on board. No soundings had been taken because the leadsman had left his post unrelieved when the watch had changed twenty minutes before. Flinders had thought they were 3 to 4 miles offshore when in fact the distance was less than two. It was a fiasco that revealed poor discipline on board and Flinders' inexperience at command. As the wind rose not long after they got off, it would've dashed *Investigator* to pieces if they'd run aground on anything but a rising tide. With no pilot and such a new, inexperienced crew, Flinders should have been on deck. Flinders blamed the chart for the incident. A later re-survey of the area confirmed that the sandbank had been correctly recorded, suggesting that Flinders did have his position wrong that evening and had gone to some length not to disclose the fact.[5]

To make matters worse, when Flinders returned to port a carpenter from another ship who'd been on board – to Flinders' knowledge only as a passenger to Portsmouth – ran off. In fact the carpenter was a deserter and had been sent on *Investigator* as an intended prisoner under guard. It seems no-one had told Flinders, so he was also in trouble for allowing a deserter to escape. Three crew members had also deserted just before *Investigator* sailed. Banks had to try to retrieve the situation; it was an extremely embarrassing position for him as patron of Flinders and promoter of the voyage.

Banks found an obvious scapegoat for the lack of discipline and its consequences. He blamed Ann. Only with her removal could Flinders have any hope of clearing up the mess. And so Ann stayed behind, and Flinders was allowed to sail on.

Flinders was very lucky. If *Investigator* had been damaged, or had Ann not been there to take the blame, he may have had considerable difficulty retaining command, despite a shortage of alternative commanders. Flinders appears to have upset others over this time, and also possibly to have had a falling out with Bass. It's hard to find evidence that Flinders was one of the officers from *Reliance* to attend Bass's wedding. Also, Elizabeth Bass, in a note to her husband, wrote of Flinders that he was:

> a man that bears a bad character … I could tell you many things that make me dislike him [–] rest assured he is no friend of yours or anyone's if another than his own interest is concerned.[6]

Whether these sentiments arose from jealousy of Flinders' past close friendship with Bass, his intense preoccupation with the *Investigator* expedition and his own career, or the way his new wife Ann had been treated will no doubt be a subject for much future speculation.

That the voyage went ahead with Flinders says something of Banks and his ability as a facilitator.

Smuggling women on ships was a bit of a tradition – but not in the British Navy. The French were quite practised at it. Bougainville's surgeon and naturalist, Commerson, was accompanied around the world by his 26-year-old 'assistant botanist', Jean(ne) Baret, and the deception was reportedly first picked up by the Tahitians. There was an officers' steward on d'Entrecasteaux's *Récherche*, Louis(e) Girondon, who was only revealed to be a woman by the Tongans. She died of dysentery on board a Dutch convoy from Batavia following the dissolution of the expedition. Louis de Freycinet would later smuggle his wife, Rose, aboard for his own round-the-world voyage.

Even if the tradition of discreet feminine presence on ships didn't run in the British Navy, this wasn't because of Banks, who wasn't unfamiliar with it. As previously mentioned, almost thirty years before, Cook's second voyage had departed without Banks after his furious withdrawal at the last minute following the debacle with the alterations to the *Resolution*. When Cook reached Madeira Island he found that a young gentleman had been waiting to join the ship as a guest of Banks but had given up and left the island just three days before Cook got there. It was fairly clear that the 'gentleman' whom the youthful Banks had arranged to accompany him on that long voyage was in fact a woman.[7]

Australian landfall, and tragedy

Flinders finally sailed from Portsmouth on 18 July 1801. After a relatively uneventful passage out, he made landfall on the west Australian coast on 6 December 1801.

The voyage from Britain had been marked by the departure of the astronomer John Crosley in Cape Town due to ill health. Flinders and his brother Samuel between themselves assumed Crosley's duties. They were skilled enough to do this

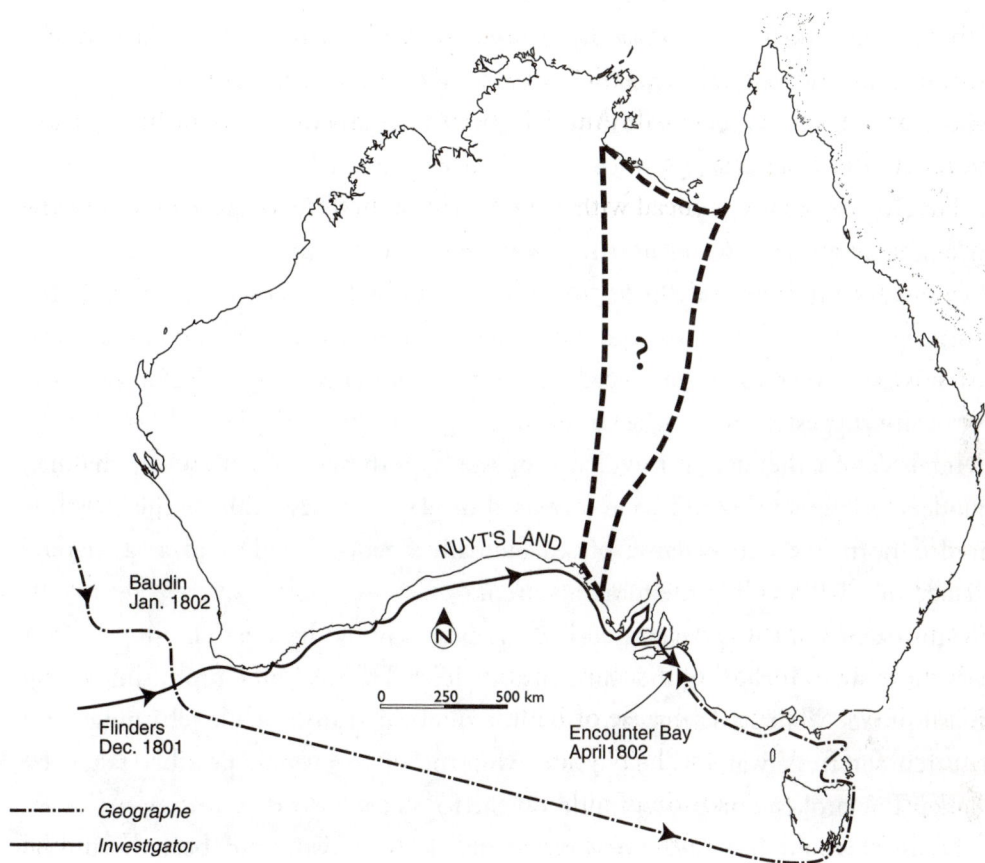

Tracks of *Géographe* and *Investigator* to Encounter Bay

and didn't want to delay the voyage while a replacement was found by Banks back in London.

After improving the existing charts of King George Sound (Albany) and the Recherche Archipelago, previously mapped by Vancouver and d'Entrecasteaux respectively, Flinders had followed the coastline to the head of the Great Australian Bight and onto coastline not before seen by European eyes, not to mention charted.

By February 1802 Flinders had only incompletely charted the coastline behind the islands of St Peter and St Francis at the apex of the Bight, hampered by shallow water and a hazy, indistinct shoreline. This was one obvious place for the suspected north–south channel dividing the mainland and linking the south coast with the Gulf of Carpentaria. Although unable to categorically answer the question, Flinders concluded, for once correctly, that the north–south channel or great inland river, if there was one, didn't empty itself there.[8] He deduced this from the specific gravity of the water nearby – when compared with water elsewhere along the coast, and the relatively little difference in salinity levels between the two – the absence of tides and the closed-in nature of the shoreline. However, Flinders himself was still not totally convinced because the presence of flocks of teal in the proximity suggested that a lake or run of fresh water existed nearby.[9]

Intensely on the alert for evidence of the hypothesised north–south channel, Flinders' excitement would have increased on 21 February 1802 as the coastline headed north into an expanse of sea (actually Spencer Gulf) and as a marked change in tidal flow became more apparent.

This excitement was devastated late the next day by disaster. The ship's cutter with eight men, including the ship's master, John Thistle, and a promising young midshipman, William Taylor, capsized in a tidal rip at dusk while returning from a search for fresh water. Their bodies were never recovered, possibly taken by sharks. The broken cutter was found on the rocks the next day.

Flinders would have felt the loss keenly. John Thistle had been bosun on *Reliance* with Flinders. He'd also been with Bass on his famous whaleboat voyage of 1798, and accompanied Bass and Flinders on both *Norfolk* voyages. Promoted quickly to midshipman and then to master, Thistle had taught himself basic astronomy and surveying. He had all the hallmarks of another Cook, and one must wonder where he would have figured in history, with these talents, had he survived. Cape Catastrophe now marks the site of the tragedy at the entrance to Spencer Gulf, with Thistle Island, as well as Taylor Island and others named after each lost crew member, standing sentinel at the approaches to Port Lincoln.

Despite this tragedy Flinders surveyed the entirety of Spencer Gulf, Gulf St Vincent and the north coast of Kangaroo Island, and concluded that a north–south channel didn't exist here.

Although these discoveries weren't insignificant, and sufficed to deliver Flinders' name to the history books, it was hardly what he'd hoped for. He'd made no further progress towards finding a maritime route inland for further continental exploration. The riddle of the rivers remained unsolved. The mouth of the nearby Murray River, one of Australia's longest, was obscured from view from the sea. Again Flinders missed a big river.

It was also here that he met Baudin, sailing along the coastline in the opposite direction to Flinders. Flinders had at least known that Baudin's expedition was somewhere on the Australian coastline, having departed Europe well before him. Baudin had never heard of Flinders. Feelings would have been intense on that April day as *Géographe* hauled wind to receive *Investigator*'s boat. It would have been clear to Baudin from the direction from which *Investigator* had come that Flinders may well have pre-empted Baudin with discoveries on that southern stretch of coast. The bay where this meeting occurred is still called Encounter Bay.

CHAPTER SIXTEEN

*F*LINDERS AND BAUDIN –
THE RACE BEGINS

The meeting between the two vessels must have had an impact on both expeditions, but more so on Baudin, who'd been longer away and had suffered many more hardships. In both ships the encounter would have destroyed the mystery and excitement – and the potential for career- and fortune-making discoveries – that would come from sailing along an unknown and uncharted coastline. Each ship had sailed over a coastline which the other had been hoping to be first discoverer. Each ship would also still have been feeling the recent loss of eight men.

As the ships passed alongside each other, Flinders called for *Géographe*'s identity. When Baudin replied 'French', Flinders asked if he wasn't speaking to Captain Baudin, to which a surprised Baudin replied that he was. Flinders lowered a boat to visit, even while keeping his cannon trained on the apparent wartime enemy, and Baudin hove to.

Flinders, accompanied by botanist Robert Brown as interpreter, was shown to Baudin's cabin, where he asked to see Baudin's passport. Having studied it, he offered Baudin his own, which Baudin didn't even read. Baudin found Flinders reserved about what he was doing on the coastline. Flinders later said that as Baudin was far more open about his own activities there, Flinders was simply happy to listen. Baudin also criticised the British chart of Bass Strait he had with him for the manner in which the north coast of Bass Strait was laid down, until Flinders pointed out the footnote which mentioned that the work had been performed by Bass from his open whaleboat. Baudin didn't appear to realise that it was Flinders' name that appeared on the chart.

As it was already late evening, the commanders agreed to eat together the following morning on *Géographe*, when Flinders presented Baudin with his most recent chart of Bass Strait, printed after Baudin had left Europe. At that meeting the French were a great deal more inquisitive, having learnt far more about *Investigator's* activities by talking to Flinders' boatmen the night before.

Flinders then explained briefly the *Investigator's* work to date, although the conversation appears to have largely been conducted in English and it seems that Baudin didn't understand much of what Flinders was saying. He did understand that Flinders had followed the entire southern coastline from Cape Leeuwin and concluded that no north–south channel existed, as a result of which Flinders later remarked that he appeared to be somewhat mortified. Flinders also told him of two anchorages he'd used at Kangaroo Island, but also, reported Baudin in his journal, that his companion ship had been separated in a gale. As Flinders had no companion ship, there was either unintentional or intentional miscommunication going on.

According to Flinders, Baudin then asked him his name once more, and only on being told it did he realise with astonishment that it was the same name as on the chart he'd criticised the night before. Brown also recorded the interview from a botanist's perspective. Baudin had shown them one of Boullanger's charts from the western coastline. Brown records Flinders' later comment that the chart was rather below mediocrity. Brown himself regarded some of their sketches of Aboriginal men and women from Van Diemen's Land as characteristic, but not well executed.

Baudin's 'somewhat mortified' appearance may have been from assessing his successes to date, which in terms of new discoveries as opposed to more detailed survey work hadn't been great. He'd discovered a couple of hundred kilometres of new coastline, with no major ports or rivers, along the southern Victorian coastline to Encounter Bay. Ironically, as the two commanders parted the following morning, neither realised that they sat only a few kilometres from the mouth of the powerful Murray River that empties into the bay.

Flinders could count more successes, including discovery of Spencer and St Vincent gulfs and Kangaroo Island; he could also claim to have disproved the existence of any north–south channel connecting this coastline with the Gulf of Carpentaria. Additionally, he could cite first discovery for the coastlines from Encounter Bay westward to the islands of St Peter and St Francis, deep in the Great Australian Bight, as well as the northern coastline of Kangaroo Island.

Original Flinders chart of south coast,
showing Thistle Island to the left and Encounter Bay, lower right

It appears that Flinders was relieved to hear that Baudin hadn't sighted King Island, at the western end of Bass Strait. It'd been seen by an earlier trading vessel and reported in Port Jackson, leaving Flinders with the potential kudos for locating and charting it on his pending journey on to Port Jackson.

Baudin in the Bight

At their meeting, Baudin noted that Flinders had been unable to clearly ascertain that there was no river system behind the islands of St Peter and St Francis. Could it still be possible that the elusive North–South Channel from the gulf country emptied itself here, not at Spencer Gulf or Gulf St Vincent? Baudin suspected this might still be the case. He would devote much time to this speculation, both on this voyage and a later voyage. Flinders, too, had resolved to return with *Lady Nelson* to settle the matter once and for all. Each must also have asked himself why he'd dallied too long at earlier anchorages – Baudin in Van Diemen's Land and Flinders at King George Sound (Albany) and the Recherche Archipelago (Esperance), and so denied their expeditions the claim of discoverer of the whole southern coastline and a greater place in history. When they met again at Port Jackson, the frustration etched in his voice, Henri de Freycinet would express it thus to Flinders:

Captain, if we had not been kept so long picking up shells and catching butterflies at Van Diemen's Land, you would not have discovered the South Coast before us.[1]

Péron was later to blame Baudin for the delays that had dogged the expedition, although unfairly so. Baudin's delay had arisen largely from the search for Boullanger and through difficulties in finding water. Much of the remaining time had been spent for Péron's benefit gathering anthropological information from the local Aborigines while charting what the French named the Northern River (the Derwent River) and Maria Island. In fact both Baudin and Flinders had devoted about the same amount of time, some two-and-a-half months, charting in Flinders' case the mainland's south-west coastline and in Baudin's case the east coast of Van Diemen's Land, before their meeting.

Flinders, discoveries and more disappointments

The two ships parted, Baudin continuing west towards Kangaroo Island and the Bight, and Flinders east to Bass Strait and Port Jackson.

Neither Flinders nor Baudin would make a crucial impact on mainland hydrography before their next meeting at Port Jackson. On his journey there Flinders landed at the north end of King Island; its existence, but not its position, had been known to sealers and to Flinders since 1799. Fickle winds prevented Flinders from making a full survey. On 26 April he then stumbled across, and made a full survey of, Port Phillip Bay, also thinking himself its first discoverer. Baudin had seen neither King Island nor Port Phillip on his journey west – despite Péron's later claim to the contrary – due to his distance offshore from each.

Flinders, however, wasn't first discoverer of Port Phillip nor, for that matter, King Island. Upon arrival at Port Jackson he discovered that Murray in the little *Lady Nelson* had pre-empted him. At King's directions, Murray had sailed *Lady Nelson* through the Rip and into the wide harbour of Port Phillip on 14 February 1802 and also had surveyed King Island, although with gross inaccuracy. Again Flinders must have regretted the five weeks spent at King George Sound and the Recherche Archipelago on the south-west coastline, consolidating the discoveries of others.

Murray's rough and far less accurate charts would be published, but Flinders' maps wouldn't be available in time for the next expedition planned from England – Lieutenant Colonel David Collins was going to settle Port Phillip.

This expedition was scheduled to leave England in early 1803, and therefore accurate charts of Bass Strait and its dangerous entrances were an imperative. For this reason Flinders had been instructed to first run down the south coast to discover if there were any new harbours. Flinders had done this, but because he'd also stopped at King George Sound and perfected Beautemps-Beaupré's and Vancouver's existing detailed charts of area, he'd wasted valuable days. For by the time he reached Port Jackson in late April 1802, he discovered that he had less than six weeks to fully document his discoveries in order to dispatch them on the next ship leaving for England – the last for a while. Unfortunately this was also the last ship likely to reach Britain in time to have the discoveries available for

Collins's expedition. Sadly for Flinders, he wasn't able to meet this timetable, being intent on producing detailed charts of the whole southern coastline, as well as the harbour at Port Phillip, as opposed to more general coastal charts to accompany the important, detailed Port Phillip charts. He sent to Banks only a report, with no accompanying charts; Banks was less than impressed.

Nor was Earl St Vincent, the First Lord of the Admiralty, who'd formed an unfavourable opinion of Flinders from his behaviour in early 1801 before his departure from Britain and reputedly had made more recent unkind remarks about Flinders' apparent indolence. The earl had little confidence that Flinders would complete the tasks assigned to him, particularly as Collins's expedition would now embark with charts no more recent than those from Flinders' *Norfolk* and Murray's more questionable work on *Lady Nelson*. In fact he subsequently offered Lieutenant James Tuckey the role of surveying the east and north-east mainland Australian coastlines once he reached New South Wales. Governor King was ordered to find an appropriate survey vessel for him. Tuckey left England with Collins in *Calcutta* and surveyed Port Phillip for the aspiring colony. His unfavourable survey was a factor that was to lead Collins to move the settlement from Port Phillip to the Derwent River, founding Hobart. After completing this task, with no survey vessel available, Tuckey left for Britain in 1805.

In later correspondence Banks quietly reminded Flinders of the less than desirable consequences of his poor time management, while accepting his blamelessness. Banks would have been acutely aware of the First Lord's poor first impressions of Flinders after the faux pas with Ann and the subsequent grounding on The Roar at Hythe.

Flinders was frustrated at narrowly missing the opportunities of first discovery along the south coast, at King Island and at Port Phillip. He may also have suspected that his delay in sending back completed charts to London might not be favourably regarded by his superiors. Following his arrival in Port Jackson on 9 May 1802, he would have been anxious to recommence his survey as soon as he could refit *Investigator*, after almost ten months at sea.

An additional reminder of the mounting French interest in his tasks also greeted his arrival at Port Jackson. *Naturaliste* was already there.

Géographe limps to Port Jackson

Baudin continued west in *Géographe* after his meeting with Flinders off the Fleurieu Peninsula. He first stopped at Kangaroo Island and next examined the then unnamed Gulf St Vincent, which he called Golfe de la Mauvaise[2] for the extreme fatigue it caused his crew – the deep-draught *Géographe* had to tack constantly in its shallows.

This fatigue would be compounded by water shortages, which particularly caused discontent among the scientists; dysentery, which still lingered from Timor; and scurvy, now breaking out on an alarming scale. *Géographe* pushed on up the Great Australian Bight as far as the islands of St Peter and St Francis and the nearby Denial Bay, as Flinders named it, adverse winds having denied him the opportunity to approach it. Baudin was, however, also denied access by these winds and was forced to leave the existence of the north–south channel as doubtful but still unconfirmed.

By the time Baudin left the coast at Denial Bay to seek water and a refit at Port Jackson, he had only thirty men – not quite half the required amount – to handle his ship, and very few for the demanding job of helmsman.[3] The strains on board showed from this issue alone. About a week earlier Henri de Freycinet, one of his senior officers, had asked Baudin to order the carpenter and the caulker to relieve the helmsmen. When Baudin suggested that Freycinet merely ask them,[4] given the differences in the tasks and the disharmony created by requiring seamen to perform jobs for which they weren't trained, Freycinet refused. He said that an officer couldn't make such an approach to men of an inferior station.

Baudin's response was blunt. He ordered not the untrained caulker or carpenter but the midshipmen to take an hour-and-a-half of each watch at the wheel. All but one of the midshipmen refused. Baudin's reaction was to refuse all the dissenting midshipmen any duties at all. He was left with a shorthanded and fatigued crew, and only one of the junior officer corps on duty, a young midshipman named, coincidentally, Charles Baudin. It was hardly a satisfactory outcome.

Baudin's shortage of firewood and water further added to his concerns, as did his desire to rest his crew and resupply his ship. Port Jackson was the closest place for this. Mistakenly he thought that winds blew constantly from the east through

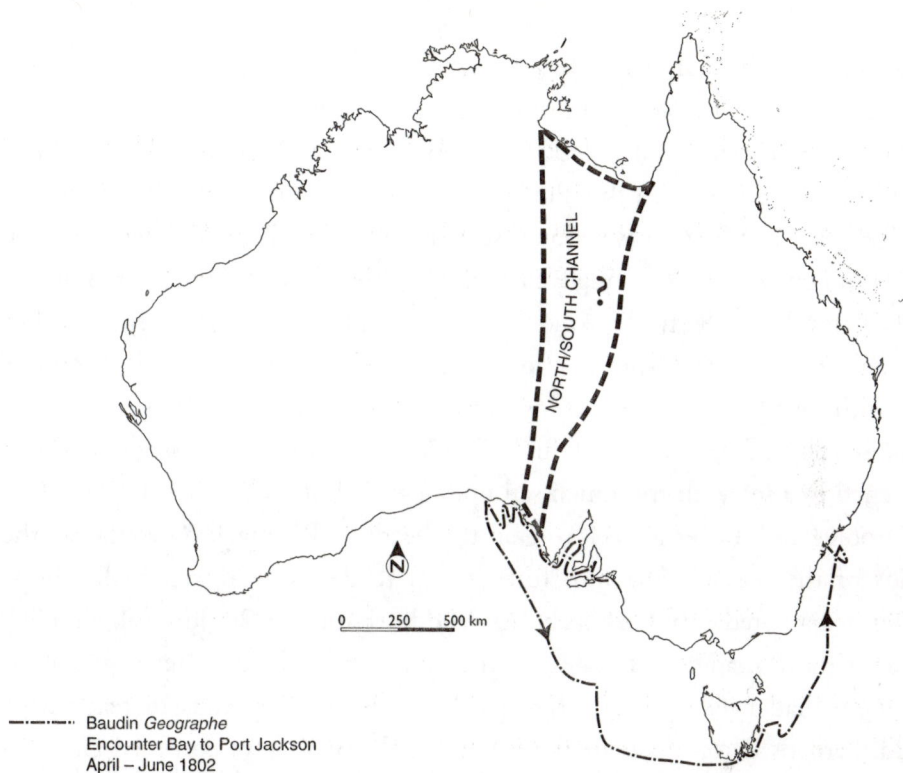

NORTH/SOUTH CHANNEL

?

N

0   250   500 km

————— Baudin *Geographe*
Encounter Bay to Port Jackson
April – June 1802

Baudin's track from Encounter Bay to Port Jackson, June 1802

Bass Strait, given the failures of so many past navigators to sail from the west. This flawed deduction put Baudin off a retreat through the little-known Bass Strait, leaving him with no option but to head to Port Jackson via the southern tip of Van Diemen's Land.

Despite the increased distance, Baudin quickly reached Adventure Bay. From there, although his crew was exhausted, he decided to further map the north-east coast from St Helens Point to Swan Island, not knowing that Boullanger had completed this while adrift in the large dinghy. As with Flinders and Murray, it would be Boullanger's labours that were recognised in the published charts.

With the growing number of sick men, all hands were required on deck to manoeuvre the ship in the squally winds that at one stage threatened to wreck *Géographe* on Schouten Island. On 3 June, changing course by wearing the ship took over half an hour, a manoeuvre normally involving a few minutes only. On an urgent call the next day to prevent anchors being lost, both watches brought forth only ten men from a total of seventy-five. With so few to con the ship Baudin finally abandoned this survey to head directly for Port Jackson.

Five days later Baudin was due east of Port Jackson but still battling contrary winds and currents and staying well offshore (some 140 kilometres) in order to do so. His crew was even more weakened, so that only twice could they wear ship to change course in one 24-hour period. Baudin didn't want to shake reefs out of his sails with easing winds. He had insufficient men to reef them back in should the wind strengthen again, and if his sails blew to shreds in a squall or the ship broached in a sea with too much sail up, he would never see Port Jackson.[5]

It took Baudin over a week to gain the heads of Port Jackson as the southerly winds of the season blew him further and further north. He finally dropped anchor, unassisted, off Port Jackson's Middle Head on 20 June 1802. History seems to have maligned Baudin at this point again; Péron later suggested that *Géographe* had needed help before passing through the heads to reach Middle Head. Perhaps Péron misremembered events. He may have confused *Géographe*'s tow under pilot the next day to the quieter waters of a mooring in Neutral Bay; this wasn't unusual with sailing ships as they progressed up Port Jackson's harbour in variable winds.

The French at Port Jackson

Now begin five months of rest and recuperation for the French in the British colony. This is of interest when considering the two countries were still at war – the shortlived Peace of Amiens wasn't reported in the colony until a short time after the French arrival – but also as a comparison to Flinders' later treatment by a French colony at Mauritius.

Baudin was lucky to have Governor Phillip Gidley King as his host. King was fluent in French and, as a lieutenant with the First Fleet, was one of the last people to dine with La Pérouse before his disappearance. Like Baudin, King found botany fascinating. He was a great correspondent with Joseph Banks. He was also interested in surveying the colony's coastlines, a subject that would similarly immerse his son, Phillip Parker King, then only ten years old. In adulthood this lad would be the next surveyor to enhance Baudin's work on the north-western coastlines. Governor King and Baudin became firm friends, and their private letters suggest much about the characters of each.

Baudin was so impressed with King's hospitality that when he sailed he left with King a number of open letters of safe passage for future British voyagers. These were addressed to all French commanders and governors. In them Baudin described the kindness shown by King to Baudin's ships and asked them to demonstrate similar concern to any bearer of the letter.

Flinders showed Baudin his recently constructed charts, which diminished considerably the discoveries of Baudin on the south coast once the work of Grant and Murray on *Lady Nelson* was included. Before then, Baudin would have assumed he'd been the first to chart the coastline from Western Port to Encounter Bay. Instead he'd charted just 200 kilometres of new coastline and missed Port Phillip as well. Baudin hadn't been aware of Grant's voyage, and perhaps for this reason he made polite excuses – bordering on fabrication – as to why his own complete charts weren't available to show Flinders. While absence of Boullanger at Encounter Bay would have been a good reason for hesitancy then, such an excuse wasn't appropriate at Port Jackson, and one must assume Baudin didn't want to reveal to the inquisitive Flinders anything about the quality of his charts.

The *Naturaliste* wasn't at Port Jackson when Baudin arrived. Although aware from Flinders' arrival in the interim that Baudin was somewhere on the south coast, she'd left over a month earlier to return to France. Port Jackson hadn't been able to resupply *Naturaliste* with all she needed due to food shortages from recent droughts. Perhaps Hamelin wasn't as capable at charming King as Baudin. His junior officers thought he no longer wished to be consort to Baudin and instead wanted an independent command. This would be understandable, given the collisions and constant separations of the vessels.

However, shortage of provisions and capricious winds in Bass Strait forced Hamelin's return to Port Jackson, where both crews were reunited only a week after Baudin's own arrival. Flinders also made several visits to both ships, and the colony warmly welcomed the explorers.

Géographe's scurvy patients were transferred to the colony's hospital, where they made much quicker recovery. Despite the current drought, cattle from the colony's vital breeding stock were slaughtered to feed them. The general feeling of good-will towards an enemy could be officially endorsed with recent arrival of news of the Treaty of Amiens between France and Britain. Péron befriended Lieutenant Colonel Paterson, a fellow of the Royal Society, who hosted expeditions for Péron into the interior. Péron also met George Bass, who'd just returned with a cargo of pork from the Pacific Islands, much of which Baudin was to purchase. Péron and Bass had something in common with their education in medicine and interest in natural history. King allowed Baudin to purchase the *Casuarina*, a 29-foot schooner that had been constructed in the colony. With her shallow draught Baudin could use *Casuarina* in much the same way as Flinders planned to use *Lady Nelson*, as a consort for shallow-water work. *Naturaliste* was to be sent back to France with all the expedition's collections to date, as well as its undesirables. In this way Baudin planned to divest himself of all the useless midshipmen he'd con-stantly complained of, including Bougainville's son, Hyacinthe.

Péron's presence at Port Jackson created headaches for both the French and the British. Péron wrote a description of Port Jackson that would form the basis for his later report to the governor of Mauritius on the colony and its strengths and weaknesses from social, economic and military perspectives. The report included a recommendation for Port Jackson's invasion by the French, a high point in grat-itude from Péron. This report would one day have severe consequences for Flinders. Péron's dialogue with Lieutenant Colonel Paterson also caused further consternation for Baudin.

The quiet rivalry between Flinders and Baudin would have only intensified in Port Jackson. Baudin wasn't the only one still harbouring some hope of a large river system or the north–south channel behind the islands of St Peter and St Francis. Flinders recorded his intention to return to properly survey this area, 'especially in the bays of the mainland', on the *Lady Nelson*.[6] One can only won-der how he felt about King making available to his French rivals a similar craft, the *Casuarina*.

Flinders had his own difficulties. He wished to resupply *Investigator* and, joined by *Lady Nelson,* continue his survey as soon as possible. He had to arrange an expe-rienced crew for the *Lady Nelson* and also replace the loss of Thistle and seven other experienced men at Cape Catastrophe on the *Investigator*. Murray he retained as commander of the *Lady Nelson*, with a midshipman, Lacy, lent from *Investigator*. The rest of the crew was made up from convicts who were promised freedom in exchange for good behaviour. Flinders found a replacement for poor Thistle in John Aken, who would later prove a most loyal subordinate through extremely trying circumstances. Aken had been the mate of the merchant ship *Hercules*, recently arrived in Port Jackson, and literally jumped ship to join *Investigator*, arriving as Flinders was weighing anchor. A replacement cutter was built, copied from Bass's famous whaleboat, preserved with almost religious devotion on slips out of water in Sydney Cove. Flinders also arranged to take with him Bongaree, the Aboriginal man who'd sailed with him in *Norfolk*, as well as a younger Aborigine, Nanbaree.

Flinders was keen to recommence his southern survey, but as it was midwinter, cold westerly winds then prevailed in the south. He opted instead to take an anti-clockwise approach and use the more promising northern conditions with the mild south-east trade winds then blowing to first examine Torres Strait, the unknown east side of the Gulf of Carpentaria and the north-western coast before the north-western monsoon struck with the wet season in November. The November monsoon would bring winds blowing from the north-west until around March the following year, the period known as the 'cyclone season'.

Baudin obviously thought differently, disclosing to Flinders his intention to head for Bass Strait and the south coast, then to proceed clockwise around the continent in the opposite direction to Flinders, meeting Flinders at the Gulf of Carpentaria in December or January 1803. Flinders believed that the cold westerly winds along the south coast would prevent Baudin reaching the Gulf before the winds in the north had swung to the south with the south-east monsoon in the following March. The March winds blowing from the southeast would hinder Baudin from approaching the Gulf from the west, Flinders reasoned.[7]

Flinders' plan, however, meant that he had to complete his survey of the Gulf before the north-west monsoon hit in November, and also trust that it wouldn't impede his passage west once it had arrived.

That two experienced navigators could differ so much on this demonstrates how little was known of the vagaries of the seasonal winds along the coastlines. That neither would achieve his goal demonstrates the impossibility of forecasting propitious winds for survey work at this early stage.

CHAPTER SEVENTEEN

FLINDERS HEADS NORTH

Flinders headed north on *Investigator* from Port Jackson on 22 July 1802. He made slow progress, sailing with the much slower *Lady Nelson* and conducting survey work on the east coast. He was looking for a river system on the north-eastern coast that would allow *Lady Nelson* to access the inland. Again he wasn't entirely complying with his orders, which stipulated an accurate charting of the unknown north-west coast and the Gulf of Carpentaria as priorities, and after that Torres Strait. Detailed charting of the east coast was far less a priority. However, Flinders was gambling for the big discovery that the south coast hadn't really produced for him.

This gamble would fail. As in the past, the major rivers would continue to elude Flinders. He made exhaustive searches at what appeared initially promising inlets at Keppel Bay (between Gladstone and Rockhampton), Shoalwater Bay (north of Yeppoon) and Broad Sound (between Mackay and Rockhampton), finding nothing.

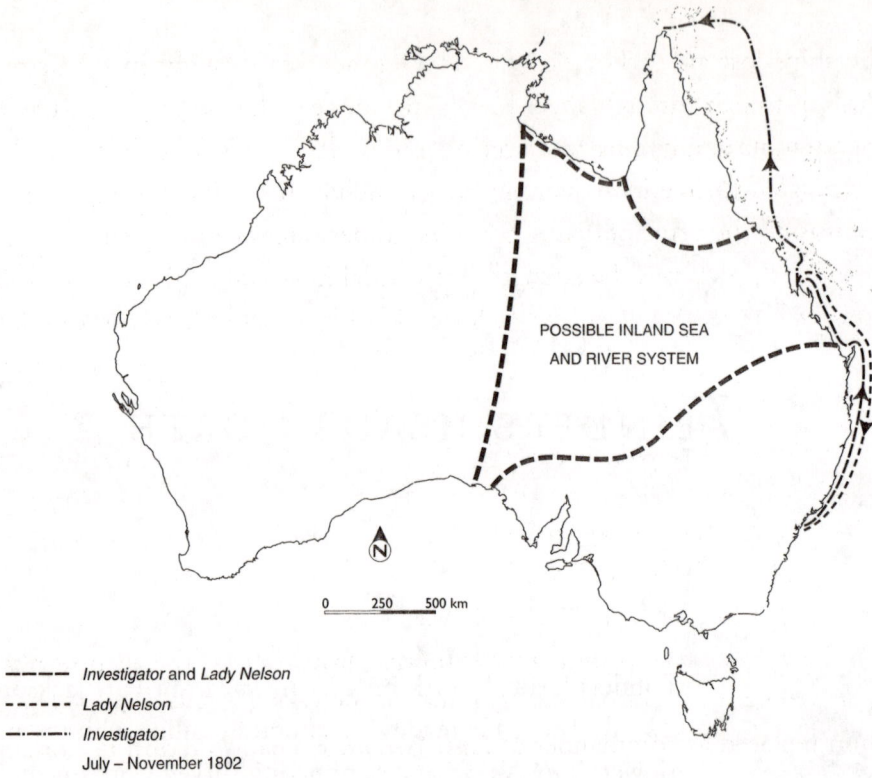

POSSIBLE INLAND SEA
AND RIVER SYSTEM

0   250   500 km

- - - - - *Investigator* and *Lady Nelson*
- - - - *Lady Nelson*
- · - · - *Investigator*
July – November 1802

Flinders' tracks to Torres Strait, November 1802

Lady Nelson is forced back

Powerful tides were hampering the small ships. Those at Strong Tide Passage, near Shoalwater Bay, swamped and swept away the new cutter while it was being hoisted, fortunately without loss of life. The copy of Bass's stout whaleboat was never seen again.

From Broad Sound Flinders headed for Torres Strait but found his course hemmed in by a reef system that appeared to have no northern end. Although not the first to experience this, he was the first to describe these reefs as 'Barrier Reefs'.[1] By 7 October Flinders was starting to worry, not only because of the danger that the coral and the tides presented to his ships, particularly *Lady Nelson*'s three vulnerable sliding keels, but also a fear that he'd wasted too much time there, and that the November monsoon would set in before he reached the Gulf.

The ships lost an anchor each on 11 October while caught in tidal streams, attempting to exit through small breaks in the reef. This deterred Flinders from future attempts at anything but large openings, despite his need for haste. As the crew struggled to free its remaining bower anchor in the fast-flowing tide, *Lady Nelson* was on the verge of diving bow-first underwater before the anchor was cut loose and lost, such was the strength of the tidal flow. After this Flinders decided to send *Lady Nelson* back to Port Jackson. With her relative lack of speed, loss of anchors – she was reduced to one undamaged bower – and damage to two of her three sliding keels – loss of the main keel, and part of the stern keel – she was more a liability than a support to *Investigator*, and they parted on 18 October 1802. On *Lady Nelson*'s return trip an anchor was manufactured from hardwood to replace her battered bowers. She returned to Port Jackson without mishap, apart from apparently grounding in soft mud next to the governor's wharf when the wooden anchor refused to sink.[2]

Sadly, *Lady Nelson*'s commander, Murray, drops out of the picture now. A bureaucratic issue in London as to whether he'd properly qualified as a lieutenant saw him replaced as commander of *Lady Nelson*. He had to return to London to clear things up. He continued with survey work on the English coastline but after 1810 he ceased active work. He was removed from the navy list, still a lieutenant, in September 1833. As effective discoverer of Port Phillip and as the first surveyor of King Island, like Grant, he deserved better. Such seemed to be the fate of most who served with the equally deserving but little recognised *Lady Nelson*. As observed earlier, the brave little vessel, after many years of devoted colonial service, disappeared with all hands north of Timor as prey to local natives in 1825. All that remains of her is what is believed to be one of her carronades, or small cannon, discovered in 1981.

Flinders threads the needle

Flinders continued to seek a large exit from the Reef. Three days after *Lady Nelson*'s departure he found his escape in the form of a wide passage out through the reefs, now known as Flinders Passage – Flinders' advice to mariners seeking this passage was poetic:

The commander who proposes to make the experiment must not, however, be one who throws his ship's head around in a hurry, so soon as breakers are announced from aloft; if he don't feel his nerves strong enough to thread the needle, as it is called, among the reefs, while he directs the steerage from the masthead, I would strongly recommend him not to approach this part of New South Wales.[3]

With the Great Barrier Reef now behind him Flinders had to move fast to make up lost ground; the imminent November monsoon could prevent him sailing through Torres Strait and surveying the Gulf of Carpentaria.

On 29 October Flinders re-entered the Reef through the same entrance used by the ill-fated *Pandora*. Over the next three days Flinders nosed his way through this shallow passage, following intermittently the paths of Cook's *Endeavour*, Bligh's voyage in the *Bounty*'s open launch and Captain Edwards's voyage both in *Pandora* then, following its wreck, in open boats with surviving *Bounty* mutineers in 1791.

Samuel Flinders took his usual careful lunar observations to correct Cook's and Bligh's various reported geographical positions. In some instances he found variations of over 100 kilometres from Cook's positions. Cook had had neither time-keeper nor, due largely to poor weather, the benefit of lunar observations, indicating how inaccurate his dead reckoning could be amid the unknown currents in the straits. Interestingly, Bligh – in his open boat, after being cast adrift from *Bounty*, with more handicaps than Cook – was more accurate in his positions than Cook.[4]

Flinders resolved to return one day with *Lady Nelson* to perform a fuller survey,[5] rather than merely to follow the tracks of others in haste because of the forthcoming monsoon. He would be back, but in an entirely different ship and set of circumstances, the first of which were about to break on him.

The enigmatic Dutch charts of the Gulf

It was with a mixture of apprehension and nostalgia that Flinders commenced his Gulf survey down its eastern coastline. The apprehension was of being caught on a lee shore with the north-west monsoon. The nostalgia was of surveying coastline that had been the first of Australia ever to be charted by Europeans – by Jansz in the small yacht *Duyfken* in 1606. Flinders surveyed one particular bay that he was sure had also been visited by the Dutch, judging from its similarity to a bay marked on the old charts. He noted the striking reddish cliffs on its low headlands, coloured by the rich bauxite deposits that have since put Albatross Bay and its mining township, Weipa, on the commercial map.

No European had been to the Gulf for over 150 years; the most recent was Tasman, who in 1644 charted what he saw – or so it was believed. No narrative had ever been published of his voyage, making the reliability and authorship of the old Dutch charts questionable. Flinders hoped to find his north–south channel emptying itself in the Gulf or at least a large river leading south to some other part of the continent, despite no trace of such a thing on the Dutch charts. It wasn't inconceivable that the old charts themselves might be hiding this.

As previously mentioned, the Dutch were renowned for deliberate inaccuracies in their charts, concealing secret discoveries for the same reasons that the Spanish had tried to keep the Torres Strait a secret for 150 years. For example, the Dutch charts marked many large rivers but Flinders could see no evidence of any. Once again, Flinders missed all the big rivers here, including the Wenlock, Watson, Archer, Holroyd, Coleman, Mitchell, Staaten, Gilbert, Nassau and Norman rivers, despite many of these appearing in some form on the charts. This would have been partly explained by his rushed schedule, which allowed only limited land exploration. Also the shallow waters required him to cruise some kilometres offshore, and a lack of suitable vantage points along the flat, low coast meant that it wasn't possible to see breaks in the shoreline.

A tone of frustration arises from his narrative following yet another search for an elusive river marked on the old charts – Van Diemen's River: 'If this place had any title to be called a river in 1644, the coast must have undergone a great alteration since that time.'[6] What he may have overlooked was the seasonal change that causes a great alteration every year, the wet season. He was there at the very end of the dry season, when the drainage from the rivers was at its lowest, assuming they drained at all at that time. The Dutch had usually come during or at the end of the monsoon – the wet season in the new year when the rivers would be in flood and more visible from the sea.

The coastline headed west. Flinders now began to harbour solid doubts about any new strait or passage; perhaps the old Dutch charts were right after all.

Decay, skulls and shipwreck

Worse news was to follow on 26 November 1802. On Sweers Island, at the south-west end of the Gulf, the crew was able at last to safely examine *Investigator* – she'd been shipping as much as half a metre of water per hour in Torres Strait and needed a thorough examination. The results were alarming. The *Investigator* was rotten, so much so that in a gale the ship would probably founder. They concluded that with continued fine weather she could run on for six months longer. Should she run aground at all she'd go to pieces in volatile conditions. Most of the decay was in her upperworks and beam ends. One view is that the decay was caused by a combination of Flinders' zealous insistence on sluicing out the upper decks each day – a practice he'd have followed from Bligh and Cook – and an

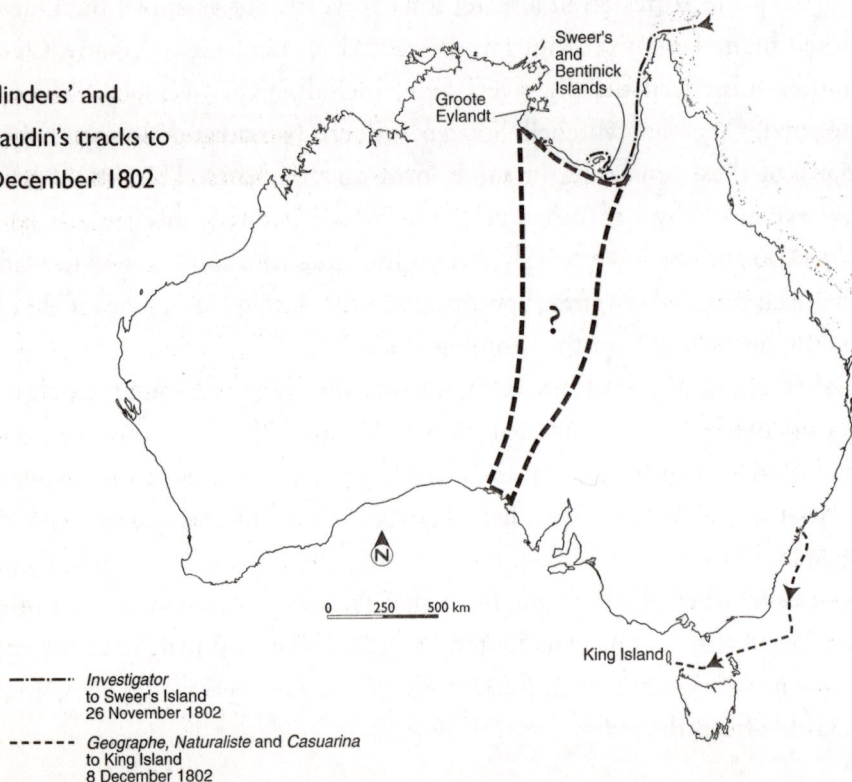

Flinders' and Baudin's tracks to December 1802

Sweer's and Bentinick Islands

Groote Eylandt

?

N

0   250   500 km

King Island

—··—··—··— *Investigator*
to Sweer's Island
26 November 1802

— — — — *Geographe*, *Naturaliste* and *Casuarina*
to King Island
8 December 1802

148

insufficiency of drainage ports allowed for by the naval contractors when originally refitting *Investigator* for this voyage. Others believe that *Investigator* was already rotten when she left England, although it would be surprising if this hadn't been noticed during her extensive refit, particularly when gun ports were cut in her sides.

Whatever the reason, there was no way Flinders could continue his planned survey in this ship. This would have galled him all the more; Baudin was due in the region any time and could take the glory of any discoveries.

One wonders in these circumstances what Flinders and his men thought of a more gruesome discovery on Sweers Island. They found seven human skulls and bones, many of which appeared to have been in the remains of three nearby fires,[7] and also part of the quarterdeck of a ship. On nearby Bentinck Island were the stumps of over twenty trees, felled with an axe, as well as broken remains of an earthen jar. Flinders concluded that an East Indiaman had been wrecked here, some of the crew had been killed by local Aborigines and the survivors had escaped on rafts constructed from timber on Bentinck Island.[8] Just the sobering thought to accompany his own crew in their crippled ship on such unfriendly shores for the marathon return voyage. They would indeed need luck.

The approaching November monsoon was less of a threat now that Flinders could sail with the protection and sea room offered by the Gulf's west coast. With the return of the south-east monsoon in March, he could hope to continue around the west coast, via Timor, and nurse the ailing vessel as quickly as possible all the way back to the closest safe haven, Port Jackson. He couldn't go back the way he'd come. The impending north-west monsoon would blow him onto the eastern shores of the Gulf, the corals of Torres Strait or the Barrier Reef while threading his way through them. At least he'd find some assistance at Timor. He'd need to hope he wouldn't encounter one of the many winter gales on the southern coast or a cyclone on the north coast. That he met neither showed that his earlier luck hadn't entirely deserted him, but he'd see no more of Australia's north-west coast.

CHAPTER EIGHTEEN

Baudin heads south

As Flinders was contemplating the tattered remains of his survey plans in the Gulf, Baudin had only just set sail. Full refitting had taken over five months in Port Jackson, and it wasn't until 17 November that his small flotilla of three ships, *Géographe*, *Naturaliste* and *Casuarina*, accompanied by an American ship, *Fanny*, cleared Sydney Heads, heading south to chart King Island and the remaining islands in Bass Strait before returning to the south and north-western coastlines. He was leaving at the time at which he'd originally predicted he would meet Flinders in the Gulf. The slow *Naturaliste* and the even more sluggish *Casuarina* would merely cause more delay.

Apart from the discovery of six convict stowaways and a collision between *Naturaliste* and *Géographe* that luckily caused no damage, leaving Port Jackson was uneventful. The collision was caused by Sub-Lieutenant Heirisson, as officer on duty on *Naturaliste*, who:

… although he considers himself a very experienced sailor, he scarcely knows how to trim his sails and has not the slightest notion of how to get himself out of an awkward spot once he has put himself in it.[1]

150

Baudin must have been looking forward to the departure of the *Naturaliste* from the expedition at King Island in eight days time. On it he was sending back to France all the officers he regarded as useless.

King Island and a British pantomime

On 8 December 1802 they were anchored off King Island at the western end of Bass Strait, on the eve of the *Naturaliste*'s departure for France, for good. *Casuarina* had been allowed the independence of surveying alone the Hunter Islands at the north-west corner of Van Diemen's Land, and one of *Géographe*'s boats was packed off for ten days to circumnavigate and survey King Island under command of Midshipman Charles Baudin. Commander Baudin dined with Hamelin before his departure. Baudin's demeanour towards his other departing officers, none of whom saw him – aside from one, and then only because Hamelin had invited him to the dinner – glares out from the ironic comment in his journal after two years' sailing together: 'As I left I thanked them for their politeness and told that I was very happy to have no farewells to make to them.'[2]

The *Naturaliste* still had a small part to play in a comic drama that would shortly unfold at King Island. Following the dinner and before *Naturaliste* sailed, the British colonial schooner *Cumberland* arrived under the command of Acting Lieutenant Robbins who, together with Surveyor Grimes, told Hamelin that Governor King had ordered him to form an advance party to help settle the Derwent River and D'Entrecasteaux Channel. He'd been blown off course to King Island by a gale that prevented him reaching the proposed settlement site. Hamelin and his officers took this news with extreme indignation. Hamelin's view was that:

> … there remained no doubt that the English are about to take from us the D'Entrecasteaux Channel, where it would however interest the French Republic very much to have a settlement, as I shall try to prove in a report I shall send to the Minister of Marine.[3]

Again it would seem that the French had been doing nothing but chasing butterflies and collecting seashells!

Nevertheless Hamelin weighed anchor and sailed the next morning for France, without further discussion with Baudin. This would have angered Baudin even more once he was delivered the same message by Robbins later that day and started to guess what was really going on. The *Cumberland* had passed a package to British Surgeon Thompson, travelling as a passenger on the *Naturaliste* back to England, addressed to the British Admiralty, containing details of Governor King's real plans, which were markedly different from what Robbins was telling everyone.

What was going on?

Nothing. The whole thing was a ruse created by King. According to *Géographe's* junior officer Ronsard's journals, it was the meddlesome Péron who'd precipitated this small flurry of subterfuge. During one of his trips to the countryside with Colonel Paterson when at Port Jackson, Péron had asserted that settlement of the D'Entrecasteaux Channel was the object of Baudin's expedition. He'd also passed a chart of the area to Paterson. Paterson only discussed this with Governor King after Baudin had sailed from Port Jackson, and only after King had heard a rumour about the chart and specifically asked Paterson what he knew about this. King therefore needed to check what was going on. There was no intention on King's part to settle the D'Entrecasteaux Channel or the Derwent at that stage, albeit that Collins did initiate a settlement at Port Phillip shortly after, which, as we've seen, would relocate to the Derwent.

Baudin claimed later that this was obvious enough from the private letter from King that Robbins also delivered to him. Baudin concluded that the *Cumberland* had been sent only to watch them. What would have annoyed Baudin was that Péron's gossip had provoked the fictional official response by King as relayed by Robbins – namely the pretence that the English were in the process of settling the D'Entrecasteaux – which was now believed as fact on board *Naturaliste* and would be reported as such to the French Government in Paris. It was all the more galling as the same ship carried the real truth and explanation in the sealed package to the British Admiralty, now in the care of Surgeon Thompson.

In one deft stroke, King produced an impression in France that the French had been pre-empted in Van Diemen's Land, an impression that his own British Government would realise from Surgeon Thompson's package was no more than a feint, and the French Government would understand only after further news from Baudin could reach France. Embarrassingly for Baudin, this confusion would be spread by a French ship under his command. And all because of Péron.

One can understand Baudin's thoughts on this man, and the irony in Baudin's words drips from his journal of the following day as he describes the departure of Péron and his fellow savants for a day ashore:

> The large dinghy also set off, carrying the scientists, their knowledge and their baggage, for these gentlemen never move without pomp and magnificence. The cooks with their utensils, the pots, the pans and the saucepans cluttered up the boat so much that not everyone could fit in.[4]

Baudin was so annoyed by all this that he stormed back to his cabin, wishing he'd sent the whole lot of them home on *Naturaliste*.

The irony was less heavy as Baudin described Robbins' and Grimes's visit later that day to entreat him for supplies, which would include sails, padlocks and even powder for their guns. Some advance party for settlement! Robbins also tried to get Baudin to replace his anchor and, when he refused, Robbins asked Baudin to repair it. Some days later Robbins and Grimes went ashore to dine with the French scientists – after asking Baudin to do further work on his anchor – but not before suspending a Union Jack over the French camp with an armed marine posted beneath. Baudin described these efforts as childish, particularly as Robbins had inadvertently hoisted the flag upside down. Baudin initially thought that it had been hung out to dry.

Baudin, a man with much humanity

A private letter Baudin sent to King, in reply to King's letter to him over these events, shows much of the humanity in the man on the subject of claiming land like King Island or the D'Entrecasteaux Channel from the Aborigines:[5]

> To my way of thinking, I have never been able to conceive that there was justice or even fairness on the part of the Europeans in seizing, in the name of their governments, a land seen for the first time, when it is inhabited by men who have not always deserved the title of savages or cannibals that has been

153

freely given them; whereas they were still only children of nature and just as little civilised as your Scotch Highlanders or our Breton peasants, who, if they don't eat their fellow men, are nevertheless just as objectionable ...

From this it appears to me that it would be more glorious for your nation, as for mine, to mould for society the inhabitants of its own country over whom it has rights, rather than wishing to occupy itself with the improvement of those who are very far removed from it by beginning with seizing the soil which belongs to them and which saw their birth ...

If you will reflect upon the conduct of the natives since the beginning of your establishment upon their territory, you will see that their aversion for you, and your customs, has been occasioned by the idea which they have formed of those who wished to live amongst them. In spite of your precautions and the punishments undergone by those of your people who have ill-treated them, they have been able to discern your projects for the future, but being too weak to resist you the fear of your arms have made them emigrate, so that the hope of seeing them mix with you is lost, and you will soon remain the peaceful possessors of their heritage, as the few who now surround you will not long exist.

Baudin wasn't blind to the commercial opportunities that Van Diemen's Land could offer France. English sealers were established in Bass Strait – Baudin dumped his six Port Jackson stowaways with them – and whalers would also be interested in these areas. Péron would later promote the potential for these industries both in the south and on the west coast, blind to the risk of over-exploitation. Baudin wasn't. His more measured evaluation of the sealing industry was also articulated in his private letter to King:

There is every sign that in a short time your sealers will have drained (King) island of its resources through the hunting of fur seals and the sea elephants. Both will soon abandon their territory to you, if they are not allowed time to replenish the losses they suffer daily from the destructive war carried on against them. They are already beginning to be much scarcer than at first and in little while you will hear that they have entirely disappeared, if you do nothing to put matters in order.

Baudin was a man well ahead of his time.

King was already trying to allocate different areas to sealing expeditions from America and France, as well as England, but this wasn't enough. Bass Strait sealers were a law unto themselves.[6] Within two years of Baudin's visit there were violent clashes between English and American sealers, largely due to the scarcity of seals, as correctly predicted by Baudin.[7]

King Island, while significant as a navigational hazard in any case, was also a place where a number of political and moral undercurrents flowing around the expedition, as well as much of Baudin's character, came to the surface.

Baudin would see little more of King Island. Bad weather forced his hurried departure – temporarily leaving his beloved scientists ashore – and the loss of another longboat on 11 December 1802 as it was being hoisted under way. He returned some days later for a planned rendezvous with *Casuarina*, fresh from charting the Hunter Isles, and to retrieve the scientists he'd watched depart some days earlier with such pomp and magnificence.

At King Island, Baudin purchased from some visiting sealers some dwarf emus – now extinct – a kangaroo and three wombats for the ship's collection before setting off to search for the overdue *Casuarina*. The sealers also used this opportunity to quietly re-stow on *Géographe* five of the six convict stowaways from Port Jackson dumped by Baudin some days before. Presumably the sealers had tired of feeding them, as well as Péron and the other briefly marooned scientists, from their own limited resources. Baudin didn't discover the stowaways for some days, by which time he was far out to sea. This cannot have improved his temper.

Not finding *Casuarina* at the Hunter Isles, Baudin again returned to King Island to leave some casks of biscuit for her and found her there; she'd lost her anchor on the treacherous bottom the day before. Together they proceeded to Kangaroo Island and then to the southern coastline to complete the chartwork done the year before by *Géographe* when she was alone and without either hydrographer. *Casuarina* carried out the survey of Gulf St Vincent and Spencer Gulf in January 1803, with instructions to return to meet Baudin at Kangaroo Island or, failing this, at the St Peter and St Francis islands in Nuyts Archipelago at the head of the Great Australian Bight. As in the past, these rendezvous would prove unworkable.

Baudin's and Freycinet's tracks to the Bight/King George Sound
(Albany) – March 1803

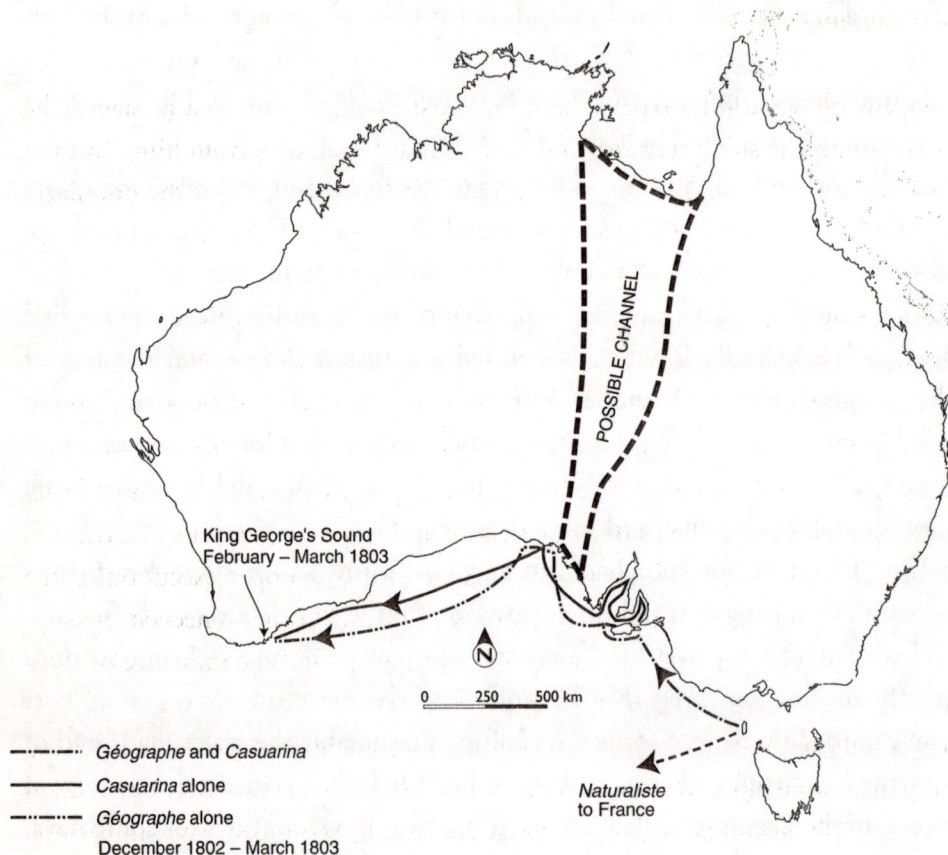

POSSIBLE CHANNEL

King George's Sound
February – March 1803

N

0   250   500 km

—·—·—· *Géographe* and *Casuarina*
———— *Casuarina* alone
—·····— *Géographe* alone
December 1802 – March 1803

Naturaliste
to France

Baudin's iciness with Péron continued at Kangaroo Island in early January 1803:

[Other scientists had returned from botanising with nothing of value] but citizen Péron was full of joy at having collected three or four molluscs, two small lizards and half a dozen ear-shells like the ones the sailors have filled their chests with.[8]

He makes similar comments about his officers observing latitude:

156

It was to the yeoman of signals that this task was entrusted, for most of the officers aboard have no instruments and those who have regard it as beneath their dignity.[9]

And on sawing wood for a replacement longboat Baudin writes that he sawed the first plank with the ship's steward, and that Ronsard took over from him, 'but the other officers were careful not to show up … Work, they say, is for the populace: a naval officer should know only how to guide a ship and to rest when his watch was finished.'[10]

So much for equality and liberty following the French Revolution.

Baudin's problem in all this was that although he was a skilled seaman and a careful navigator, this was a hydrographic mission, and he – unlike Cook, Bligh, Flinders and other English commanders – had no surveying skills. His predicament was similar to that of the luckless Lieutenant Grant, the first commander of *Lady Nelson*, who also lacked surveying skills. Baudin was forced to rely on the two expedition hydrographers, Faure and Boullanger, for the proper execution of the work. For the navigation of the ship, although skilled himself, he needed the support of his junior officers, which, due to their young age and the influence of civilians like Péron, he didn't have.

Péron had caused Baudin angst not only because of his involvement in the King Island pantomime but also because of his innate ability to get lost ashore and delay departures. It's also likely that Péron, the young wounded veteran of the Republican Wars, didn't have a high opinion of Baudin on principle because of the latter's past in service of the Austrians over that time, so was an additional voice against him among the young officers.

The expedition was also a scientific mission, but Baudin, although a keen amateur botanist and collector, was no scientist. In this he was no different from the English navigators, who also relied on the services of external experts, but to nothing like the same degree. Many of the twenty-two scientific personnel with whom Baudin started the expedition were vocal, relatively immature, ambitious and unused to submitting to anything too undemocratic in the early days of the new French republic, especially the strict autocracy of a naval ship. La Pérouse had had five such staff and d'Entrecasteaux six. Cook had had six, and Flinders, on *Investigator*, five.

Considering Baudin's numerous pressures, it's understandable that in order to bring the expedition any success, his health would pay a heavy price.

The south coast charted, and more tensions

Louis de Freycinet, commander of *Casuarina* and brother to Henri de Freycinet of the shells and butterflies remark, completed the surveys of the western coasts of Spencer and St Vincent gulfs but arrived a day overdue at the Kangaroo Island rendezvous, by which time Baudin was already under sail. Again the breakdown of relations and communication between Baudin and Freycinet off Kangaroo Island led to separation of the vessels. Baudin was already under way on opposite tack but furled his mainsail to slow down – but not stop – for *Casuarina*'s expected tack and overhaul. This didn't happen, and *Casuarina* just continued on her way as though Baudin wasn't there.

Perhaps after over two weeks of freedom Freycinet wasn't ready to play consort again. His excuse was that he'd tacked but expected Baudin to heave to and wait, an unusual comment for a subordinate to make of his commander, the latter in a much larger vessel. Baudin wanted to clear nearby islands and not stop until then. As things turned out, by the time Freycinet did tack, *Géographe* was out of sight, and the ships wouldn't meet again for some weeks, passing each other again in the night when Baudin finally did heave to.

In the meantime Baudin completed his final task for this coastline, to see if there really was a river or other passage behind the elusive islands of St Peter and St Francis at the top of the Bight. South-easterly winds, which Flinders had believed wouldn't assist Baudin, continued to prevail, much to Baudin's delight.[11] He was able to settle at last that these islands offered no nautical access to the interior, something that no navigator to date had been able to do conclusively. Denial Bay would live up to its name in full. There was no north–south channel here.

Baudin then headed for King George Sound, where he finally caught up with the *Casuarina*. Freycinet, short of water and too far west, had by-passed Denial Bay and continued on to King George Sound. Again the difficulties of managing consorts was demonstrated. Baudin saw that *Casuarina* was supplied with sufficient rations to reach Mauritius; he clearly no longer needed her. He also believed that Freycinet would take the first opportunity to head for Mauritius,[12] and by now must have been exhausted and frustrated by the constant separations. Furthermore, he thought that should *Casuarina* discover anything, it wouldn't be reported to him anyway.[13] As it turned out, most of the remaining survey would be fit only for a vessel as shallow-bottomed as *Casuarina*.

April 1803

0   250   500 km

------- *Geographe* and *Casuarina*
——— *Casuarina* alone
------- *Geographe* alone
March 1803 – April 1803

Baudin's and Freycinet's tracks to Timor, April 1803

The west coast, Timor and termination

Baudin's substantive plans were to head north and complete the survey of both the northern coastline and the Gulf of Carpentaria. Now it was the seasons, rather than illness or personal frictions, that would disturb these plans. It was March 1803. The north-west monsoon that had so worried Flinders was soon to finish, and the south-east monsoon was likely to start. These winds would make south-

and east-bound coastal work, particularly in the Gulf, almost impossible. It's hard to guess how Baudin could have achieved his aims as he had therefore only less than a month to get from Albany to the Gulf of Carpentaria in order to do so.

A stop at Shark Bay brought on more worries for Baudin in the irresponsible antics of Péron. Péron had persuaded artist Petit and gardener Guichenot to come with him on a day's shell collecting and exploring. Péron's total lack of any sense of direction had landed him in dire straits before, and it happened again. After two days out without food or water they staggered back to the boats, Guichenot still faithfully carrying some of the treasure trove of shells Péron claimed to have found. Guichenot had achieved no botanising and Petit no sketching. They'd also had a frighteningly close shave with an aggressive band of Aboriginal people. They managed to hold them at bay by pointing a totally defective firearm at the locals and advancing three abreast with no protection other than deception and brazen courage. Luckily for them the Aborigines retired. This was the third time Péron had become lost and overdue, and Baudin now directed that their 'learned naturalist', as Baudin referred to him in his journal, only be allowed ashore if Baudin himself was with the shore party.[14]

As to Shark Bay itself, Baudin suggested that a whaling enterprise could be very successful. French interest would remain in this area, as well as Swan River and King George Sound, well into the nineteenth century. Unfortunately, Baudin's chart of this coastline, particularly from North West Cape to the Dampier Archipelago, was run from some distance offshore. Consequently, offlying islands were continually mistaken for the coastline, for instance Barrow Island, lying over 50 kilometres offshore. This wouldn't be corrected for almost fifty years, following John Lort Stokes's survey in *Beagle* in 1841. *Casuarina* could have assisted with independent inshore survey work, but Baudin had lost faith in Freycinet's ability to get to a rendezvous on time or communicate results.

Nevertheless, Baudin was able to construct a much more accurate picture of the coast from Port Dampier almost to current-day Darwin than indicated by charts from Tasman's voyage almost 160 years before – these had been the only other charts available. Baudin conjectured that the Dutch charts were intentionally deceptive,[15] just as Flinders had in the Gulf. Baudin wrote in his journal that the Dutch charts generally seemed wrong, that both latitude and longitude seemed intentionally altered to suit Dutch convenience, as the Dutch couldn't but have avoided some of the reefs, rock and other dangers that Baudin encountered, but these weren't marked on the charts.

Baudin lost an anchor on 18 April, leaving him with only three and still much of the coast to chart. *Casuarina* now only had one anchor from her original five,[16] to add to his worries.

Baudin's health was declining fast. His tuberculosis, initially contracted at Timor in 1801, resurfaced in the form of a persistent, dry cough. The lemon juice he'd purchased at Port Jackson was served to the crew to alleviate thirst in the hot conditions.[17] It would also, unbeknown to him, keep scurvy at bay. Like d'Entrecasteaux before him, the commander was the sickest man on the ship. Lack of water forced the expedition to break from the survey again at Timor in late April 1803.

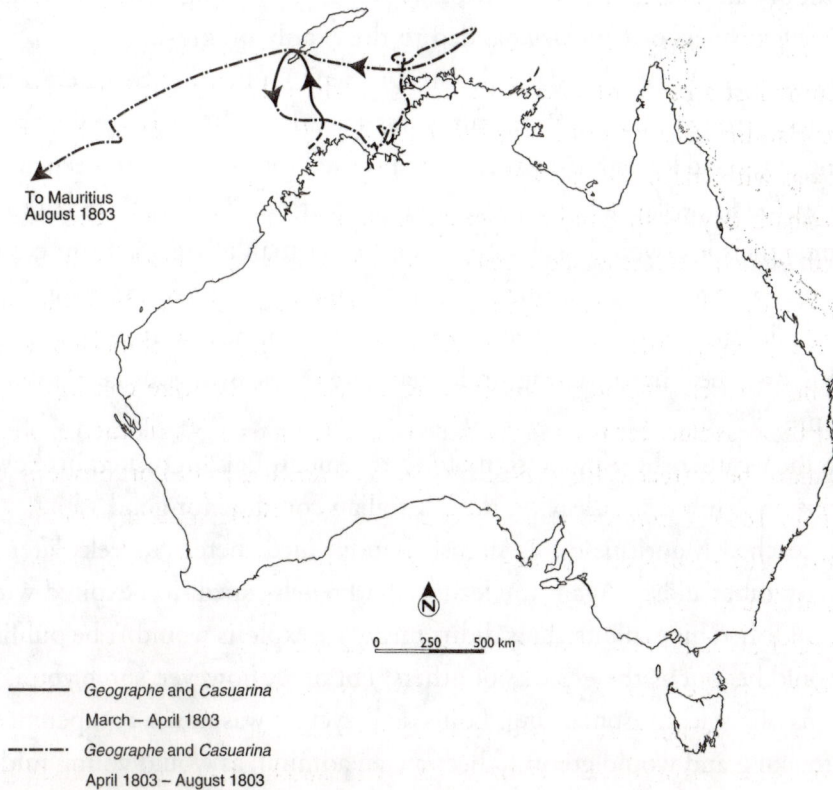

To Mauritius
August 1803

0   250   500 km

N

—————— *Geographe* and *Casuarina*
March – April 1803
— · — · — *Geographe* and *Casuarina*
April 1803 – August 1803

Baudin's and Freycinet's tracks after Timor

After Timor, Baudin returned to the north-western coastline. Timor's legacies of dysentery and venereal disease again reappeared in the crew. Bernier, the astronomer, died from dysentery on 6 June 1803. Baudin was now starting to spit blood together with, as he conjectured calmly in his journal, pieces of lung.[18] The south-east trade winds had arrived, and Baudin was advised that these winds wouldn't allow him to continue his survey, just as Flinders had foretold. Baudin was convinced from his experience on the south coast that the inshore winds would be benign, despite contrary offshore winds. This time his conviction was wrong. Winds remained light and variable initially, but by the end of June tended to blow from the south-east as he'd been warned. The low coastline needed to be passed close up – it wasn't visible at 6 kilometres or more offshore. However, with the variable wind threatening to blow the ships onto the at times shallow shore-line, Baudin had no alternative but to stand out to sea. Strong currents to the west and the sluggishness of *Casuarina* added to their tardy progress.

By 27 June 1803 Baudin sighted Van Diemen's Gulf, close to where Darwin now stands. Strong buffeting from winds was starting to threaten the lives of their surviving emus and kangaroos, which they'd started force-feeding to keep alive. As Baudin would know from his last triumphant return to Paris in 1798 with exotic live plants, their loss would probably be far greater than any gains from complet-ing the survey in these conditions.[19] They also had no more than a month's sup-ply of biscuit and two months' water, with twenty men now ill from Timor afflic-tions, his own health worsening and, again from his own journal, 'nobody to replace me'.[20]

With the weather showing no sign of improvement, Baudin turned his bows for Mauritius on 7 July 1803, leaving the Australian coastline for good.

They reached Mauritius on 7 August. Baudin died there six weeks later from pulmonary tuberculosis. Again the leader of a French expedition expired without having published his exploits. This didn't mean the exploits wouldn't be published. They would be, but as the exploits of others. For once, however, some young gen-tlemen may have learnt something. Louis de Freycinet was now a competent hand at chartmaking and would go on to become an admiral, as would young midship-man Charles Baudin. Freycinet would later lead his own expedition to the Pacific.

CHAPTER NINETEEN

*F*LINDERS' LUCK RUNS OUT

As the crews of the *Géographe* and *Casuarina* were enjoying the tropical delights of Mauritius after over two years' absence from French territory, Flinders and his crew were contemplating a wholly different fate, sitting as castaways on a coral atoll, surrounded by the wreckage of their ship. When we last saw them at Sweers Island in the Gulf ten months before, this image was nothing more than a fear. At that time Flinders and his crew had just found that *Investigator* was rotten; they were surrounded by the remains of another luckless ship, the bones of its crew and hostile local inhabitants; there were thousands of kilometres along the west coast of Australia between them and safety. Yet the wreckage presently surrounding the castaways was not of the *Investigator*, and they were sitting on the Great Barrier Reef off the Australian east coast, not its west.

Aboriginal people – art, bloodshed and benevolence

Following discovery of the rotten state of *Investigator*'s timbers at Sweers Island in late November 1802, Flinders completed his survey of the western coast of the Gulf of Carpentaria, taking great care to ensure that his ship didn't ground

163

inshore. She nevertheless grounded a number of times off Mornington Island in the Wellesley Group, which Flinders later named.

The fact that Samuel had forgotten to wind the chronometers that day as well must have further tested his temper. This would mean either a forced stop for astronomical observations required to rerate the chronometers, continuing with potentially inaccurate measurements of longitude until rerating was more convenient (an unlikely option), or otherwise dispensing with their use entirely (not an option at all). Flinders chose the second option.

As he headed further north, Flinders was finding more traces of Asian visitors – the bêche-de-mer (trepang, or sea slug) fishermen who come from the north with the favourable monsoon winds. Evidence of their occasional occupation convinced Flinders that his earlier deductions about castaways at Sweers Island had been premature.[1] This conviction may have been strengthened with discovery of further human skeletons, in this case the result of Aboriginal funerary rites.[2]

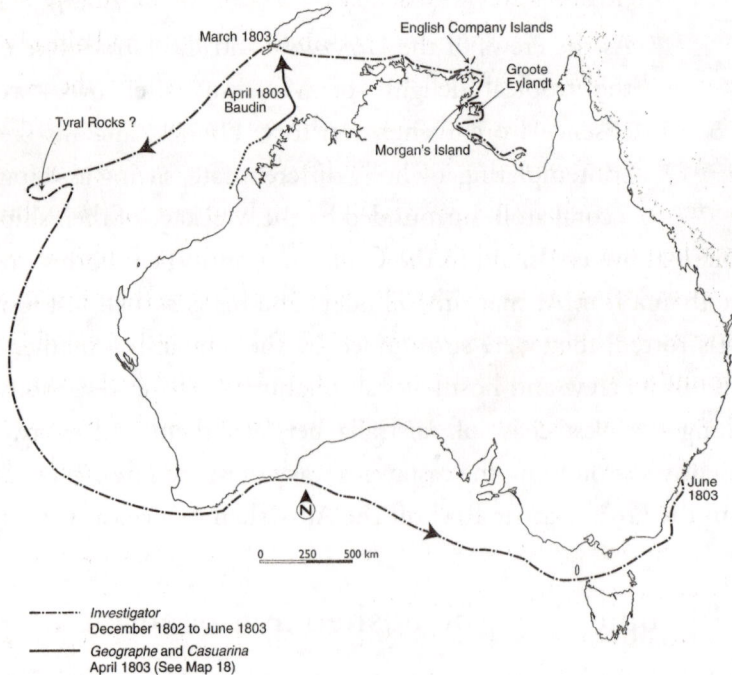

Flinders' Track – Mornington Island to Timor and Port Jackson,
December 1802–June 1803, compared to Baudin's and Freycinet's tracks

By now the wet season had arrived, consequently many of the previously dry rivers were now in flood. Even so, as Flinders coasted the west side of the Gulf he still missed the large rivers, particularly the Roper River, which by New Year 1803 would probably have been in full flood. Flinders noticed the resultant current but assumed this was tidal. Shallow water would prevent closer examination. Baudin had been right to keep *Casuarina* for his proposed survey here. Flinders could have used the services of *Lady Nelson*.

Unlike the French at Depuch Island, Flinders' men did come across impressive Aboriginal art on cliff walls at Chasm Island near Groote Island. William Westall sketched these. They found nutmeg trees growing in the cliff gullies, but of low quality compared to the Dutch nutmegs,[3] as well as a new type of fruit, *Eugenia Jambos*. This would have been of interest to Banks and the Board of Trade.

On an uninhabited island nearby, Flinders' men had a fierce clash with a group of Aboriginal men. Master's Mate Whitewood and mineralogist Allen were collecting firewood when they were approached by a party of six Aborigines who'd paddled over by canoe from a neighbouring island. The party had already unsuccessfully pursued other members of the expedition. Three of the Aborigines approached the unarmed Allen, laid down their spears and exchanged a green bough with him, a sign of peace. Whitewood wasn't so lucky. He'd kept his musket with him; one of the three Aborigines near him held out his spear, but when Whitewood reached out to take it, the man ran the spear into Whitewood's chest. Whitewood's musket misfired, and both he and Allen took to their heels, followed by flights of spears. Although none hit Allen, five more hit Whitewood before he was able to wound one of his attackers with his musket and force their retreat.[4] Fortunately Whitewood recovered, although a marine called Thomas Morgan died from sunstroke that night, having lost his hat elsewhere on the island during the excitement. Flinders named this island after him.

Sailors looking for the attacking Aborigines' canoe that evening surprised three Aborigines as they were leaving the islands. In pursuit the sailors shot one dead and mortally wounded another as they tried to paddle away. Westall sketched a dead body, found on the shoreline the following morning. The body was dissected and the head preserved in spirits.[5]

Flinders observed that the Aboriginal people of this locality were more aggressive than elsewhere and that they'd probably pursued Whitewood's crew to this otherwise uninhabited small island to fight with them.[6] He surmised that this might be due to past differences with the Macassan fishermen. Nevertheless he was anxious to avoid bloodshed and stated as much in his journals.[7]

Restraint on Flinders' part was more apparent at his next landfall, Caledon Bay on the mainland. He sent Lieutenant Fowler to communicate with some Aboriginal men on the shore that they sought water. Referring to them in his narrative as 'Australians', probably the first time the word was used, Flinders wrote:

> They staid to receive him, without showing that timidity so usual with the Australians; and after a friendly intercourse in which mutual presents were made, Mr Fowler returned with the information that fresh water was plentiful.[8]

The same Aborigines later stole a hatchet and, worse still, a musket from Brown's botanising party. The musket was returned broken the next day, on promise of another hatchet. More thieving – this time an axe – led Flinders to resort to Cook's old measure of taking hostages, the habit that led to Cook's death, to encourage the return of the missing item. In response, instead of the missing axe the Aborigines offered a young girl to Bongaree but only, surmised Flinders in his official account, for the purpose of seizing him by way of retaliation. Flinders released the hostage, a lad named Woga, three days later, despite having not had his axe returned. He saw no justice in detaining the youth further, plus he recognised the potential for later reprisals by the tribe against Baudin, expected on the coastline any day.[9]

Flinders observed that the Aborigines from the west side of the Gulf were circumcised, unlike elsewhere in Australia.[10] He speculated, but with doubt,[11] that this practice might have been inherited from contact with the Muslim bêche-de-mer fishermen, like their apparent familiarity with the use of firearms.

With his survey of the Gulf of Carpentaria concluded at Cape Wilberforce (just past present-day Gove), Flinders recorded that:

> [The coastline marked in the ancient Dutch charts] isn't very erroneous, which proves it to have been the result of a real examination; but as no par-

ticulars were known of the discovery of the south and western parts, not even the name of the author, though opinion ascribed it with some reason to Tasman, so the chart was considered as little better than a representation of fairyland, and didn't obtain the credit which it was now proved to have merited.[12]

The ancient Dutch charts had been right all along.

A new industry for Banks

In the lee of a long chain of islands off Cape Wilberforce, Flinders came across the explanation for the scattered evidence of foreigners in the Gulf, so far evidenced only by shipwreck and skulls. Up until then, Flinders had only suspected the activity of pirates in the area. The real answer came in the form of a detachment of sixty vessels from the Malay bêche-de-mer fishing fleet – about 1500 sailors – in command of a Malay, Pobassoo. The fleet used to sail in from Sulawesi with the November monsoon and return with the south-east monsoon in March. Through Flinders' Malay cook the various pieces of earthenware, shipwreck and bamboo seen in the Gulf, but not the skulls or the circumcision, were patiently explained by Pobassoo.[13] Disputes between the Malays and the Australian Aborigines were also evident. Pobassoo warned Flinders of local native treachery, having himself been speared in the knee in the past.

Flinders also obtained detailed information from Pobassoo on the collection, cure and sale of sea slug in the lucrative Chinese aphrodisiac market. Presumably he considered that it might become of interest to English commerce and, no doubt, to Banks and Hawkesbury. Sure enough, he named the long chain of islands the English Company's Islands.

At Arnhem Bay – so named by Flinders because of its similarity to a bay charted on the ancient Dutch maps, possibly by Willem van Coolsteerdt in the Dutch yacht *Arnhem* in 1623 – Flinders decided to terminate the survey. It would have been a difficult decision. The survey was becoming more interesting, and there was increasing availability of safe anchorages, wood and – perhaps not unexpectedly at the end of the wet season – water, much more so than for the earlier parts of the survey. However, it was early March. It'd been three months since Master Aken had assessed the remaining six-month lifespan of the rotten *Investigator*, and they still had the south coast gales to contend with in order to get back to Port Jackson. By now scurvy had reared its ugly head with the crew. Flinders himself could no longer climb the masthead or make trips in boats – an integral part of survey work – due to scorbutic sores on his feet. The crew were all also suffering from exhaustion from heat and humidity.

Flinders' character again pokes through the official narrative here:

… and as the whole of the surveying department rested upon me, our further stay was without one of its principal objects.[14]

This hardly recognised the enormous contributions of his brother Samuel, particularly his astronomical work in the absence of astronomer Crosley, or his faithful supporters Fowler and Aken, also so essential to the success of the survey work. Even in the official narrative they are little mentioned in this regard. Flinders was determined to monopolise all credit for the voyage, like Bligh before him.

Flinders examined the Wessel Islands; again he took the name from the old Dutch charts – it commemorates the yacht *Wesel* in Gerrit Pool's squadron which visited New Guinea, but not Australia, in 1636. The Wessel Islands guard the north of Arnhem Bay, and for Flinders were the last of the north coast. On 6 March 1803 he left for Timor and refreshment, in readiness for the final voyage to Port Jackson.

There is still evidence that his mind wasn't made up on whether or not he was terminating the survey. In his narrative Flinders only speaks of giving up returning to the north Australian coastline because of being unable to obtain salt provisions at Timor or to find a passage to send charts and dispatches back to Port Jackson. He was still not committed to return. He spent two additional weeks searching unsuccessfully for the Tryal Rocks, on which the English ship *Tryal* had been wrecked in 1622. They were ultimately located in 1818.

Port Jackson, and the same, awful legacy of Timor

Flinders' commitment to further survey changed as disease, contracted as usual in Timor, reappeared. Dysentery started to take its toll; two men died at Recherche Archipelago. By then fourteen others were extremely sick. By Bass Strait one more, the marine sergeant, died and eighteen were sick. It was the end of May 1803, and Flinders decided to curtail further exploration of Bass Strait and head for Port Jackson before more lives were lost. He didn't make the same mistake as Baudin had almost a year before by insisting on further surveying with a disease-ridden crew: 'I carried all possible sail, day and night, making such observations only as could be done without causing delay.'[15] No doubt a belief that the bottom was about to fall out of the rotten *Investigator* was additional incentive.

By the time they arrived at Port Jackson on 9 June they'd lost two more men. Four more would die in hospital ashore, including the Banks-appointed gardener, Peter Good. Good had been a tireless collector of plants and appears to have been well liked by all. Botanist Robert Brown in particular would have felt his loss, and this may been one reason why he would later complain that Flinders hadn't allowed sufficient opportunity to collect as he'd wished, although this doesn't appear to have been a complaint of Good's. As Good's own journal records, it would seem Good had plenty of opportunities to go ashore, and he appears to have been competent at finding his way in the bush – after a shaky start – a characteristic that doesn't appear to have been strong in Brown, like Péron.

Investigator was fully surveyed in port. She was most rotten in the bows and midships, all above the waterline. Flinders wrote that thirteen timbers in one area were so rotten that a cane could be thrust through them, and he was generally considered lucky not to have hit foul weather on their return voyage – it would have sent her to the bottom.

Flinders was now without a ship and still with a survey to complete.

CHAPTER TWENTY

TWO REPLACEMENT SHIPS

Governor King offered Flinders a number of alternatives to replace *Investigator*, but the only vessels immediately available were *Lady Nelson*, returned with her wooden anchor in tow the preceding year, and *Francis*, which Flinders knew well from his earlier voyages to the south. King had also offered Flinders the use of the roomy supply ship, *Porpoise*, but this would've required considerable refit for his task, costing Flinders time that he wasn't prepared to spend. Flinders opted to return to England in *Porpoise* and procure a new vessel from the Admiralty. History would play out a repetition of these circumstances sixteen years later when King's son, Phillip Parker King, landed at Port Jackson without a ship, needing to complete a survey and finding only the ageing *Lady Nelson* available.

Thirty-eight individuals from *Investigator* were taken to man *Porpoise*, including Fowler as commander. Twenty-one others, among them the Flinders brothers, joined as supernumeraries, with only nine of the original crew remaining in Sydney. Flinders didn't command, planning instead to concentrate on completing his charts during the voyage, although he could direct Fowler on survey work in Torres Strait during that time, under orders given to Fowler by Governor King.[1] Flinders left behind the Banks-appointed naturalist, Robert Brown, and the botanical artist, Ferdinand Bauer, with much of the expedition's botanical collection. Brown was disappointed with the voyage. He didn't consider that Flinders had provided him with sufficient means to preserve specimens and, as mentioned before, thought that opportunities to collect had also been reduced. The results of the expedition, therefore, were in his view unimpressive:

> In zoology our acquisitions have been extremely few; no new quadrupeds and but a few nondescript birds have been found. Ichthyology, which would have been afforded the greatest number of new species, I have been unable to attend to. Insects and shells are neither numerous nor interesting. Mineralogy has been uniformly a barren field, and even botany has fallen short of my expectations.[2]

This provides an interesting contrast to the claims of Péron as to the thousands of unknown new species delivered back to Paris by the Baudin expedition. According to Museum of Natural History director Jussieu, of Péron's collection, over 2542 species in zoology and 640 in botany were found to be new to science.[3] With more than 180 000 specimens, the Baudin expedition doubled the holdings of the museum but, as we shall see, its benefits were never fully realised.

In early August 1803, as Baudin's ships were limping into Mauritius, *Porpoise*, in company with merchant ships *Cato* and *Bridgewater*, left Port Jackson. On 17 August 1803, at night and well out to sea, the lookout on *Porpoise* reported breakers ahead in the darkness. Although the helm was quickly put down, *Porpoise*, sailing under only double-reefed topsails, didn't have enough sail to respond and she quickly went up onto the coral. In the darkness *Cato* did likewise and went to pieces quite rapidly. *Bridgewater* was able to haul off in time. In fact *Cato* went onto the reef in order to avoid colliding with *Bridgewater*, which didn't stay longer than the following morning and rendered no assistance. Instead she drifted downwind and finally sailed on to India via Batavia to report the total loss with all hands of both ships.

Wreck
Reef
+

------ Tracks of *Cato, Porpoise, Hope,*
Rolla, Cumberland and *Francis* to and
from Wreck Reef,
August and September 1803

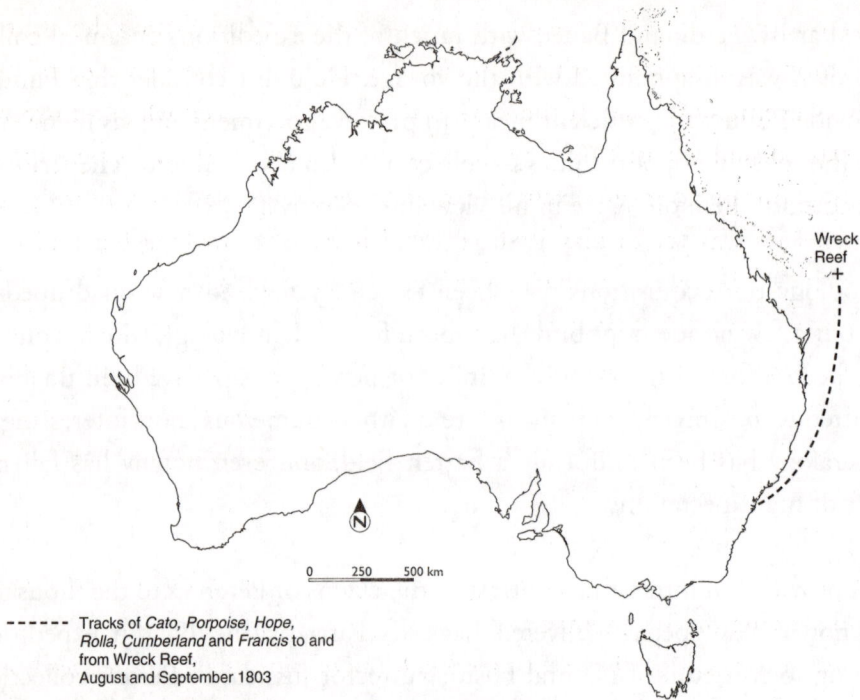

Flinders' track with *Cato, Porpoise, Bridgewater* and others (*Hope, Rolla,*
Francis and *Cumberland*) between Port Jackson and Wreck Reef

Shipwreck

Almost all hands from both wrecks swam or rowed to a small coral cay nearby.
Flinders took charge of the castaways, but only after *Bridgewater* had sailed away.
Until then Flinders had been waiting for her return in one of the boats, keeping
safe all his charts and documents – apart from the chart on which he'd been working at the time, which had been washed out the cabin window. He wanted to leave
these precious works with *Bridgewater* and then guide her to a safe anchorage.
Bridgewater, with a draught of at least 18 feet, and by morning some 8 kilometres
downwind of the wrecks, would've had a difficult task beating back into the wind
to approach shallow and dangerous waters. She wouldn't have been able to tell
from that distance that there were survivors. The cay where the survivors had gathered was only 2 metres high and unlikely to be visible from so far away. The wrecks
themselves would also by then have been deserted, both of men and boats.

Flinders' reaction – sitting in the gig with his beloved charts awaiting the assumed return of *Bridgewater* rather than immediately erecting some signal of distress, which was eventually done on the second day after the wreck when an inverted ensign was hoisted – wasn't helping the situation. In this context it's perhaps less astonishing, but by no means excusable, that *Bridgewater* sailed away bearing such bad news. Fate was yet to play its final hand here. After making her erroneous report in India, *Bridgewater* sailed on and was never seen again.

With *Bridgewater*'s departure the survivors knew they had only themselves for salvation. Nine days later, on 26 August, Flinders and thirteen others set out for Port Jackson and help in one of the two surviving ship's cutters, now named *Hope*, leaving eighty others on the cay. Flinders now faced his own open-boat voyage, perhaps not as arduous as those of Bass, Bligh or Edwards before him or Shackleton after, but important nevertheless for those stranded on the cay. Luckily he had a smooth trip to Port Jackson, covering the 1100 kilometres in thirteen days and arriving on 8 September.

Flinders was mindful of the southerly gales that might delay or destroy them. He ordered that in his absence the survivors build two decked schooners from the wreckage, sufficient in size to carry all but the remaining cutter's crew. These were to leave the cay if help hadn't arrived in two months, with the remaining cutter following two weeks after, charged with delivery of the precious charts and journals, which were left in Fowler's care. They were to be the very last thing to leave the cay.

The reef on which the ships had been wrecked wasn't charted; Flinders named it Wreck Reef and suggested also that the surrounding sea be called the Corallian Sea for the number of similar dangerous features.[4] This seems the first time that today's Coral Sea was so referred to. Evidence of past shipwreck on the cay, in the form of a worm-eaten sternpost or spar, caused Flinders to wonder whether La Pérouse had met his end on these reefs or others nearby. At this stage they would have identified with those castaways, just as they had back at Sweers Island following their gruesome find of human remains and shipwreck.

Discipline was maintained during Flinders' absence as though on board ship, with the routine reading of the Articles of War and the flogging of offenders as well as the daily task of food collection and ship-building. The artist William

Westall records some frustration with young midshipman John Franklin, who drove a herd of sheep from the wreck across some of Westall's paintings as they lay in the sun to dry. Although Westall probably didn't appreciate this, such youthful pranks would have eased tensions in other quarters.

Cumberland, and Flinders' desperation to reach England

While this activity took place on Wreck Reef, Flinders was frantically organising rescue back at Port Jackson. In order to rescue the castaways and to take those so wishing on to Canton, King made the merchant ship *Rolla* available, along with the smaller *Francis*, which was to bring others, as well as salvageable stores, back to Port Jackson. He also acceded to Flinders' desire to push on to England as soon as possible with his charts and journals. As Flinders wanted to complete further survey work on the north coast, King organised for him to take the small schooner *Cumberland*. He acknowledged in a letter to Nepean that this was necessary to 'enable [Flinders] to add to his survey and arrive with his charts etc before any other accounts reach England'.[5] Obviously both men were concerned that the results of Baudin's expedition would precede Flinders' to the printers.

The *Cumberland*, the vessel selected to preserve England's status in this different sort of 'space race', was the same vessel that had carried Robbins to King Island during Baudin's stay there, in order to discourage French settlement. The little schooner was slow, and too cramped to allow Flinders to work on his charts as he had on *Porpoise*. But Flinders' ambition and haste motivated him beyond good sense, although he didn't expressly acknowledge quite the same reasons as King had for these actions. He would save some three to four months, he estimated, on *Cumberland*, and 'joined to some ambition of being the first to undertake so long a voyage in such a small vessel, and a desire to put an early stop to the account which Captain Palmer would give of our total loss', he accepted King's offer of this 'strong, good little sea boat'.[6]

Whether the months he saved taking *Cumberland* through Torres Strait would be compensated by the time he lost not working on his charts, as he could have done as a supernumerary on *Rolla*, is debatable. Flinders also had little idea of

Cumberland's sailing qualities, although these became clear enough once they'd left Port Jackson – *Cumberland* was anything but a strong, good little sea boat. She was poorly constructed and had equally poor sailing qualities. Moreover her planking wasn't watertight and her bilge pumps were almost totally useless.[7] Flinders wrote during the voyage to Wreck Reef that if pumping ceased for an hour-and-a-half, water would wash over the cabin floor and he could only write while perched on a cabin locker with his knees for a table.[8]

The justifications given for the voyage, to be the first to make it in such a small vessel and also to pre-empt *Bridgewater*'s Captain Palmer's erroneous report of their loss at Wreck Reef, were extraordinary. These were hardly proper reasons to risk his charts, journals and the lives of good men in a further survey in dangerous waters, where he knew fresh water would be scarce until the wet season in late November. This is only the more extraordinary once he realised just how unseaworthy *Cumberland* was. Such was the nature of his desperation to precede Baudin to the printing presses. His cherished promotion to post-captain, as promised by Earl Spencer in 1801, also hinged upon his return to England.

Before Flinders left, King directed him to deliver King's own dispatches to the Admiralty and the secretary of state for the Colonies. This little detail would figure prominently later.

Given the size of *Cumberland*, Flinders knew he'd require many stops on his return voyage to England. King cautioned him against Mauritius, even though the fragile peace still existed between England and France. Sadly for Flinders, what King didn't give him were any of the letters for safe passage written and left with him by Baudin.

The rescue flotilla left Port Jackson on 21 September 1803, five days after Baudin's death at Mauritius and four days before news reached Mauritius that the Treaty of Amiens had collapsed and that France and Britain were again at war, news that wouldn't reach Port Jackson for a further two months.

They reached Wreck Reef on 7 October. Flinders recorded this as one of the happiest moments of his life. Samuel Flinders, in charge of the cay while Fowler was away in one of the boats, appeared less excited. He was in his tent calculating some lunar distances when he was informed by an excited young midshipman that ships were on the horizon.

After a little consideration Mr. Flinders said he supposed it was his brother come back, and asked if the vessels were near? He was answered, not yet; upon which he desired to be informed when they should reach the anchorage, and very calmly resumed his calculations.[9]

His brother's narrative then goes on to relate how, once Samuel had been informed that the anchorage had been reached, he ordered the salute to be fired, 'and took part in the general satisfaction'.[10]

The expedition divides

The castaways were now divided into those who wished to go on to China in *Rolla*, those who wished to return to Port Jackson in *Francis*, and the ten officers and men to whom Flinders offered the opportunity to continue the survey in *Cumberland*, despite its clearly demonstrated unseaworthiness. On 11 October 1803 the various vessels departed Wreck Reef – *Cumberland* for Torres Strait; *Francis* for Port Jackson, together with *Resource*, the little vessel constructed by the castaways while Flinders had been absent seeking help; and *Rolla* for China.

The expedition had finally broken up. Most of its members, including Fowler, Samuel Flinders, Franklin, artist Westall and mineralogist Allen – and some completed charts of the east and north coast – left on *Rolla*, and the balance, other than Flinders' small band on *Cumberland*, headed for Port Jackson.

Samuel Flinders, Fowler and the others on *Rolla*, *Resource* and *Francis* will drop from our story, but not before an epilogue that ranks with the best of ripping yarns. *Rolla* reached Canton without further mishap, and the expeditioners transferred to the next ship available for passage to England. This was part of the China Fleet, a convoy of thirty East Indiamen and other vessels wishing to make the voyage. It had become a more dangerous prospect because of the recent dissolution of the Treaty of Amiens and the resumption of war between France and England, as represented by a squadron of five French ships under Rear-Admiral Linois then cruising in the locality. In these circumstances the arrival of men from the *Investigator* and the *Porpoise*, all well-trained Royal Navy seamen, would have been welcomed with open arms by the often poorly manned China Fleet. They departed Macao in late January 1804 but ran into Linois's squadron of superior-strength ships at the eastern end of the Malacca Straits in mid-February.

Fowler and others assisted in a daring ruse that deterred the French from battle. Flying Royal Naval insignia from the masthead of the East India Company's armed ships, utilising smart, well-disciplined ship management and accurate gunnery, they adopted an aggressive strategy of direct confrontation, giving all the hallmarks of crack Royal Naval ships of the line, rather than far less threatening armed East Indiamen. The well-trained men from *Investigator* and *Porpoise* would've been needed to do this. As a result, Linois discontinued the action and allowed the convoy to escape.

Linois's own position was difficult. While the temptation to capture a rich China convoy would've been strong, he was conservative. His crews were poorly trained and he couldn't afford the loss of even one ship to weaken his squadron, now so remote from France and any source of replacement or reinforcement. The East India Company was grateful for Fowler's efforts, as well as those of Samuel and other personnel from *Investigator*. Fowler received a reward of 300 guineas and a sword valued at 50 guineas in gratitude. Samuel received a sword as well as a reward of between £125 and £150, a not inconsiderable sum.

The China convoy reached England with no further mishap in August 1804, and Flinders would have read of the heroics early the following year. His feelings wouldn't have been totally of elation, but more of frustration and, possibly, envy, given what had happened to him by then.

Following the departure of the *Rolla* at Wreck Reef, Flinders sailed for Torres Strait. This took him through waters strewn with reefs that he neither knew of or saw. Luck still had an eye out for him. Admittedly he'd learnt from the *Porpoise* wreck to be additionally cautious at night and to make preparations for quick manoeuvring if required. After Torres Strait, however, it was as though fortune really ceased to favour Flinders. He did find a new passage through Torres Strait on this trip, the Cumberland Passage. He also ascertained that one of the landmarks to the western portal of Torres Strait, Booby Island, had been wrongly determined by Cook, due to the absence of a chronometer.

Flinders touched at the Wessel Islands one last time, his last Australian landfall ever, and departed for Timor.

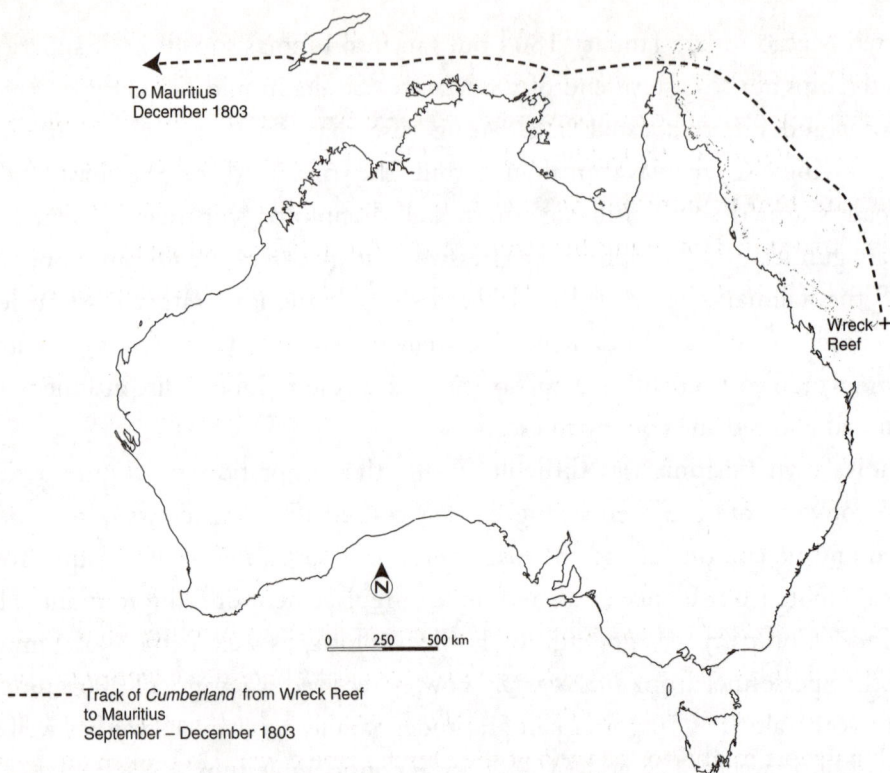

To Mauritius
December 1803

Wreck +
Reef

0   250   500 km

- - - - - Track of *Cumberland* from Wreck Reef
to Mauritius
September – December 1803

0

Flinders' track from Wreck Reef, towards Mauritius, December 1803

Despite the poor sailing qualities of *Cumberland*, Flinders gained West Timor only four days later than *Bridgewater* reached Batavia.[11] This showed how much quicker a passage from Port Jackson west through Torres Strait could be. *Bridgewater* took four months to reach Batavia from Port Jackson. *Cumberland* took less than half that time to get to Timor. The huge benefits of time saved by the Torres Strait route, a matter of months rather than weeks, were obvious, if a safe passage could be found.

However, Flinders never really addressed the high priority that exploration of Torres Strait had in his instructions from the Admiralty. He'd spent no more than five days in two visits charting the region. The Cumberland Passage (now called Flinders Passage) proved to be even more intricate than the passage he'd used for his previous visit and has never since been seriously utilised by commercial shipping. At least a century would pass, as would a number of survey vessels, before safe passages were fully charted through this maze of shoals and reefs.

Flinders reached Timor in November 1803 to find that much-needed pitch to caulk *Cumberland*'s badly leaking sides couldn't be found, and that there were no facilities for him to bore out his useless bilge pumps. He also received news of Baudin's arrival in Timor and his return to the Australian coastline in the preceding June. Perhaps it was this news that hurried Finders on ... he only stayed five days in Timor before setting out for the Cape of Good Hope.

Unfortunately *Cumberland*'s leaks became worse, with only one pump, though constantly manned, having any effect at all. Should the wind shift from port to starboard, Flinders doubted *Cumberland* would stay afloat as the only working pump wouldn't reach water until the hold was half full. Flinders decided to head for Mauritius on 6 December 1803. Aware of the possibility of blame later, Flinders recorded in his journal his reasons for stopping there. In addition to the issue of *Cumberland*'s seaworthiness, Flinders cited the possibility of obtaining passage on a faster ship so he could work on his charts, sending *Cumberland* back to Port Jackson with trade goods or otherwise selling her, plus concerns that his French passport might not be valid at the Dutch Cape if war had broken out again – he assumed it would be valid at Mauritius. There was one last reason:

... considering the proximity of Mauritius to the western coasts of Terra Australis, which remained to be examined, I was desirous to see in what state it had been left by the Revolution, and to gain a practical knowledge of the port and periodical winds, with a view to its being used in the future part of my voyage as a place of refitting and refreshment, for which Port Jackson was at an inconvenient distance.[12]

This innocuous enough entry would have harmful repercussions, as would Flinders' attempts to summons a pilot as he approached the island on 15 December 1803, having no chart. As he drew nearer to the island, he first flew a French flag at the foretop masthead, believing this to be the signal for a pilot, as well as British colours. He then followed a small schooner along the coast and into a little port, thinking that it had offered piloting services, despite the schooner's displaying a constant desire to avoid contact with *Cumberland*. He was even more amazed to see the schooner's crew anchor:

> … without furling the sails they went hastily on shore in a canoe, and made the best of their way up a steep hill, one of them with a trunk on his shoulder. They were met by a person who, from the plume in his hat, appeared to be an officer, and presently we saw several men with muskets on top of the hill.[13]

It was only once he was anchored that Flinders discovered that Britain and France were now at war again. Flinders' concerns were compounded by his passport, which he couldn't read as it was in French; it'd been briefly explained to him in 1801 before his departure from England but he hadn't looked at it since and, crucially, it only applied to the *Investigator*, not the *Cumberland*. He was informed, however, that Baudin's expedition was still in Port Louis, down the coast.

Although advised that Baudin was now dead, Flinders felt sure that others in the expedition would vouch for him, notwithstanding his questionable passport. With this in mind, with a French pilot on board, he sailed to Port Louis, less than a day's sail away. He took with him King's passport-breaching dispatches to the Admiralty and the secretary for the Colonies, still aboard intact.

$\mathcal{M}$AURITIUS, A POWDERKEG FOR FLINDERS

The Baudin expedition had been at Port Louis since August 1803. Despite Baudin's death the expedition was still enmeshed in turmoil, particularly as to the question of who should assume command. The logical choice was Henri de Freycinet, Louis de Freycinet's brother and Baudin's appointed successor. However, sitting at Mauritius was Milius, a senior officer who'd been left behind by Hamelin due to illness and who now claimed the right of command.

One other far-reaching change occurred on 17 August 1803, ten days after Baudin reached Mauritius, in the form of the arrival of 34-year-old Charles Mathieu Isadore Decaen. He was Mauritius's new governor.

Decaen

Decaen had joined the Revolutionary French Army aged twenty-two and was a zealous republican. His skill, intelligence and courage had led to rapid promotion, and he attained the rank of general by the age of thirty-one. He had the ear of Napoléon, who appointed him captain general of all French possessions in the Indian Ocean. The Treaty of Amiens of 1802 provided for the return to the French of all Indian holdings seized by the British.

Napoléon sent Decaen to take possession of these and to re-establish French influence in India while the fragile treaty lasted. Decaen had asked for the appointment, so keen was he to pursue these aims, which he saw as a worthy pinnacle in an already impressive military career. Decaen left France with a squadron of ships commanded by Linois and a small force of troops. They reached Pondicherry in India in mid-June 1803 to find that the British still hadn't vacated the French settlements, in contravention of the treaty. The English, under Lord Wellesley, the future Duke of Wellington, believed that as the treaty was unlikely to hold, there was no point in complying with its terms in any hurry only to then have to fight to take back the territory they'd handed over.

Decaen was subsequently advised by Paris that war was imminent – in fact it had begun again in May – and that he shouldn't land with his small force but should instead proceed to Mauritius as new governor. Mauritius was a remote island that'd been in French hands for some time and had formed an important base for privateer activity against English shipping from India during the previous war. It would do the same again and become a significant thorn in the side of British dominance of the East Indies and Indian shipping routes until its ultimate capture in 1810.

Decaen clearly had ambitions for greater things, and Mauritius would be an important base for achieving these ambitions.

Péron's machinations

The new governor had the highest political connections, great military skill, an impressive record and, importantly, a desire to throw the English out of India.

No doubt with this in mind, Péron presented Decaen with a report he'd prepared of the British colony at Port Jackson and its weak defences, which Péron urged the French Government to destroy. Péron went on to claim, incorrectly, that Napoléon's real object in sanctioning the Baudin expedition had been that it spy on the British, but this had had to be kept secret in order not to invalidate their passports. Decaen's experience as an intelligence officer familiar with the reports of spies, and possibly more cynical of them than ever following his recent disastrous expedition to Pondicherry, would have enabled him to filter out of Péron's rhetoric the flimsy argument supporting an invasion of Port Jackson. Furthermore, from King's dispatches on *Cumberland* he would shortly become aware of fortifications under construction there that might defeat any invasion.

Much of Péron's character is exposed in his report. Péron boasts of being received by Governor King, of being treated as a son by the commander of the New South Wales Corps, Lieutenant Colonel Paterson, and otherwise knowing all the important people of the colony, who'd unsuspectingly furnished him with information 'as valuable as it is new'.[1] This is no doubt true, and it seems those colonials were less than discreet about their local industry – just as Péron had reportedly been about the objects of the French on King Island. Péron waxed lyrical about the likely commercial success of merino sheep, sealing, whaling, Indian hemp and wine in the colony. He also claimed that perfect security reigned everywhere in the colony, and that he wasn't aware of one case of murder since the colony's formation.[2] In his view the Irish convicts would happily revolt at the least provocation, a tidbit that he proffered presumably to spur on Decaen to further action. The report, while accurate in some respects, was hardly a fitting thanks for the efforts to which King and Paterson in particular and the colony in general had gone to refreshing Baudin's sick and starving crew.

But, more specifically, Péron's report also mentioned Flinders in the context of eastern expansion by the English towards the American colonies of the French ally, Spain, and occupation of bases in the Pacific from which to harass them:

> If among the numerous archipelagos that are visited constantly some formidable military position is found, England will occupy it and, becoming a nearer neighbour to the rich Spanish possessions, will menace them impatiently. Mr. Flinders, in an expedition of discovery which is calculated to last five years, and who doubtless at the present moment is traversing the region under discussion, appears to have that object particularly in view.[3]

Péron wouldn't have been aware that as he was writing this Flinders wasn't traversing the region under discussion half a world away. He was actually only a day's sail away from, and fast approaching, Mauritius.

Péron delivered the report to Decaen three days before he sailed from Mauritius's Port Louis for France in *Géographe* on 16 December 1803. The very next day Flinders arrived at Port Louis in mistaken expectation of seeing members of *Géographe* there to vouch for him. Instead Flinders would meet a new governor, who'd recently had his own career ambitions thwarted by the British in India, in apparent breach of a written treaty, and who'd quite literally only just read a report naming Flinders as an instrument of England's political expansionary policy.

To make things worse, Flinders knew nothing of all this. He would've had his eye solely on getting his charts to England for publication before Péron – now known to be only a day ahead of him. This was critical to his own career.

If this wasn't bad enough, Flinders didn't fully appreciate how fragile his position was. He was sailing into an enemy port, in a boat entirely inconsistent with and unfit for the nature of his expedition, without a single botanist, naturalist or artist on board, none of their collections or paintings – which had returned on *Rolla* – without a valid passport, and carrying confidential dispatches which he hadn't thought to toss over the side, in breach of the passport's terms.

Flinders would need tact and diplomacy to work his way around this situation. Unfortunately he would demonstrate precious little of either. He and his crew would suffer the consequences.

CHAPTER TWENTY-TWO

*F*LINDERS' NEMESIS, DECAEN

Immediately upon his arrival in Port Louis on 17 December 1803, Flinders was taken to see Decaen, who kept him waiting some time. While waiting in the shade outside, some French officers chatted amicably with him in broken English, asking him whether he knew of 'Monsieur Flinedare' and his voyages in New Holland, 'of which, to their surprise I knew nothing, but afterwards found it to be my own name they so pronounced'.[1] His reply would also have been reported to Decaen before his meeting with Flinders.

The one disastrous interview

Flinders started things poorly with Decaen by refusing to remove his hat, a discourtesy of the time, particularly from a man already under suspicion as a spy. Decaen wanted to know why Flinders had arrived in *Cumberland* rather than *Investigator*, and Flinders stated his erroneous view that the passport was to protect the expedition, regardless of which ship carried it. Decaen wasn't convinced by this, given the express wording of the passport. He also refused to believe that Governor King would send Flinders off in a vessel as tiny as *Cumberland*, which was also a command more appropriate to a junior lieutenant than a commander. Hence there was clear doubt in Decaen's mind that Flinders was who he claimed to be, and Flinders' guarded replies earlier to Decaen's officers would've only made things worse.

Flinders was sent back to his ship and ordered to deliver up all charts and documents, including King's dispatches and his own journals. This would have further tried his fragile patience; he was still fatigued from an earlier bout of malaria and the reappearance of scurvy. He told the French officers accompanying him that unless Decaen's conduct improved he, Flinders, would neither see him again nor come ashore. When he was straightaway informed that he was to be taken ashore, he exploded at the idea of being a prisoner. He was obviously still seething and undoubtedly fairly exhausted the next day following an afternoon's further questioning by Decaen's minions. At 5 pm he was invited to dine with Decaen, but Flinders refused, saying that he'd already dined. When pressed by the French officer bearing the invitation to at least appear at the table, which would have been courteous – the invitation is believed to have come from Decaen's wife – Flinders again declined, replying:

> under my present situation and treatment, it was impossible; when they should be changed, – when I should be at liberty, if his excellency thought it proper to invite me, I should be flattered by it, and accept his invitation with pleasure. This answer being carried, the aide de camp told me, the general would invite me when I should be at liberty.[2]

Flinders says in his narrative that he believed the invitation to be a trick:

> It had indeed the air of an experiment, to ascertain whether I really was a commander in the British Navy; and had the invitation been accepted without explanation or a change of treatment, an inference might have been drawn that the charge of imposture was well founded; but in any case, having been grossly insulted both in my private and public character, I could not debase the situation I had the honour to hold by a tacit submission.[3]

In other words, Flinders believed that to accept Decaen's dinner invitation, even in these circumstances, would have confirmed him to be a spy, such behaviour being so out of character for a British naval officer.

That Flinders still believed this when he wrote his narrative around ten years later speaks volumes for the extent of his arrogance. Frustrated, disappointed and exhausted though he may have been at the time, his explanation a decade on says little for his recognising, with the benefit of hindsight, the tenuous position he'd been in and much for the suggestion that despite his charting brilliance he was arrogant and overly proud, with a view of his own importance that was much higher than anyone else's, and with little ability to appreciate the differing views of others. It may have been this quality that had given Earl St Vincent a poor impression of Flinders three years before when *Investigator* was due to sail and caused the disquiet in Admiralty circles as to the youth and inexperience of her new commander.

It's true that Flinders appeared to be a popular and able commander. Of course *Investigator*'s officers were either kinsmen, such as his brother Samuel or cousin John Franklin, or from the same locality – Robert Fowler was also from Lincolnshire. However, Flinders had never really had to interact with others on equal terms since his appointment to *Investigator*. With the powerful Banks as his patron, Flinders must have been less inclined to concern himself with the thoughts or objections of others while in British territory. On board *Investigator* he had the full authority and discipline of the navy to back up his orders. At Port Jackson, Governor King – also dependent on Banks's support – was sympathetic to all his needs. It was only at Mauritius that Flinders had to deal with people for whom these connections meant nothing. Flinders may also not have realised just how powerful Decaen was, despite his apparent youth – he was only five years older than Flinders. He was no petty provincial governor but a man Napoléon had intended would take and rule India. Like Flinders, it was only bad timing that had placed Decaen in Mauritius. That Flinders failed to adjust to this at the time, and even ten years later – at which time peace had been restored with France following Napoléon's exile to Elba – suggests character traits such as immaturity, stubbornness and insensitivity, with perhaps a little too much ambition mixed in.

The next day Decaen went through Flinders' journal and would have seen the entry in it stating, among the reasons why Flinders had headed for Mauritius, that he wished to record winds, weather, the port and present state of the French colony and its usefulness to Port Jackson. This would have confirmed in Decaen's mind that his suspicions of Flinders as a spy might have been well founded.

Flinders would never get to dine with Decaen or, for that matter, see him again. Three days after their meeting Flinders sent him a strongly worded note and four days later another, even stronger, wildly worded and accusing Decaen of ungentle-manlike behaviour. As a result Decaen refused to correspond with him, 'since [Flinders] knew so little how to preserve the rules of decorum'.[4]

Flinders would be detained at Mauritius until June 1810 – six-and-a-half years – largely as a consequence of Decaen's impressions from their first and only inter-view and supervening events.

The six-and-a-half years

Given Flinders' obsession with the delivery of his charts and himself to England before the Baudin expedition could publish its results, his detention must have been soul destroying. He refused to accept he was a prisoner-of-war and thus was-n't available for prisoner exchange – a system by which warring nations swapped back captured officers of equal rank. To do anything else, he reasoned, would be to deny his immunity under his passport – and also his chances of promotion. Naturally he brought to bear all his influence by way of correspondence to Banks, the Admiralty and the British contacts in India to resolve his predicament. Local landowners sympathetic to his position also wrote to influential contacts in Paris in support. Decaen also sought to cover his position by requesting directions from Decrès, the French minister of Marine, in Paris.

After Flinders had been in detainment almost two-and-a-half years, Banks and others finally managed to have his case heard in Paris on 21 March 1806, argued by Bougainville himself. The outcome was that Napoléon signed an order approv-ing the return to Flinders of both his liberty and the *Cumberland*. It took another year for a copy of this order to reach Mauritius, in July 1807, by which time the British had retaken the Cape, Trafalgar had been won, the French Navy destroyed, and Decaen's hopes of a French invasion of India quashed.

Ironically, this worked against Flinders. With the maritime war now swinging against the French, Decaen realised that blockade and a British naval-borne attack on Mauritius were both likely. He thought, quite rightly, that the British knew little of the island's winds, currents and coastline, therefore it wasn't in his interests to release Flinders, who by then would have gained this knowledge that could be so useful to an invading force. Further, Napoléon's order hadn't been expressly to release Flinders from the island, although it approved this course of action. In other words, it could be read that if Decaen wanted to release Flinders, Napoléon agreed. With this background as justification, Decaen detained Flinders for a further three years, continuing to regard him as a dangerous man.

The dangerous man occupied himself as much as possible to fight off depression and recurring illness and to keep his excellent mind active. He immersed himself in producing his charts, enabling Master Aken to take almost the entire series of finished charts with him for delivery to Sir Joseph Banks when he was released from captivity in May 1805. These included a general chart of all Australia as well as detailed coastal sheets. Flinders also sent a memoir on the charts, his *Book of Bearings*, containing recorded sightings from the voyage, and the first volume of his logbooks. Flinders clearly hoped that this would be sufficient to allow work to commence on engraving and publishing the charts.

However Sir Joseph merely forwarded them on to the Hydrographic Office, where they sat untouched along with later dispatches from Flinders. The ageing hydrographer, 69-year-old Alexander Dalrymple, the original promoter of Cook's *Endeavour* voyage, was now suffering from failing health and facilities, and lax administrative procedures. In this context he was desperately overworked, dealing with the hydrographic demands of a country at war and facing seaborne invasion from the French with inadequate surveys of its own shores. One might speculate that Dalrymple had neither the time nor the capability to work his way through Flinders' memoir and other correspondence to properly supervise production of an accurate chart of a coastline half a world away that no-one saw as the highest priority, other than an imprisoned lowly commander in Mauritius.

Dalrymple would be displaced from his office in early 1808 and die shortly after. His successor, Captain Thomas Hurd, wouldn't focus on the charts until after Flinders' return to London, when they met in 1811.

Usually charts were published with an official narrative of the voyage, but Flinders had never discussed this aspect with Banks, particularly the question of who'd write the narrative and, more importantly, who'd pay for its publication. The Admiralty might bear the expenses for publication of the charts, but the narrative was another matter. Unfortunately Flinders devoted very little time to this on Mauritius, either by expanding out the material in his logbooks or otherwise writing a rough draft – as Bligh had done in the past and Péron would do in the future – in case it was needed. He did write a fuller account of the wreck of *Porpoise* and *Cato*, and of his own incarceration, assuming wrongly that the Admiralty might be interested in their publication. This and other large writing activities were in any case curtailed by the British blockade of the island, which ultimately made paper impossible to procure for these purposes.

Over this period he also wrote a paper on how magnetism in ships was affecting their compasses, which was published by the Royal Society, and an account of approximate longitudes of the Australian coastline. Flinders also kept his mind active by producing further papers on applied mathematics and trigonometry, cartographic projection, magnetic dip and estimations of distance by the motion of sound. In late 1809 he wrote an endearing but uncharacteristic piece about Trim, his cat. Trim had been born on his luckiest ship, *Reliance*, and had remained with Flinders right up until that time. It may have been grief from the cat's recent disappearance that made him write. Flinders believed that the rotund and singular Trim had ended up on some starving local's plate, a casualty of the British blockade.

Flinders didn't languish in a stinking jail during his long detention. In fact he was fairly well housed for probably all but the first three months, which were spent at the small and rather dirty Café Marengo. After that he was removed to the Maison Despaux, a large country house set in more than 2 hectares of garden, where other captured naval officers were held. After over a year there he went to live with a family inland on the D'Arifat plantation, where he remained until his release, learning French from the family and teaching their daughters English. Although having given his parole to go no further than 12 kilometres from the plantation, he became close friends with nearby families. Toussaint Antoine de Chazal, one of those friends, painted one of only two surviving portraits of Flinders over that period.

Release finally came for Flinders in July 1810, as Mauritius was about to be taken by the British. Decaen let him go on the promise that he wouldn't serve against France for the remainder of the war. At Cape Town on the voyage back to Britain, Flinders was interrogated by Vice-Admiral Bertie on Mauritius's defences. Bertie was planning the invasion of Mauritius, and the information he extracted from Flinders was probably in breach of Flinders' parole. This vindicated Decaen's original reasons for detaining him after Napoléon's order allowing his release.

Flinders' unwavering conviction was that Decaen's refusal to release him was the cornerstone of a huge conspiracy designed to allow the French to beat him to the publishers. This assumes far more faith in Péron by Decaen than is justifiable. Even so, by the time of Flinders' release Péron had still not managed to publish a single chart, although he'd published an official narrative and an atlas containing drawings but no maritime charts. Given the value of published charts, if Flinders' conspiracy theory had been true, Decaen would have detained him until the charts of Boullanger, Faure and Freycinet had been published, which wasn't until 1812. Decaen was a professional soldier, with little interest in such things.

It's more likely that Decaen initially thought Flinders might be a spy, but after Flinders' churlish behaviour wasn't prepared to give him the benefit of the doubt, leaving the decision in the hands of those in Paris to whom he'd referred the case. Once this decision had been made and Flinders' bona fides were clearly established, his long stay in Mauritius and the knowledge he consequently built up of its coastline at a time when the British were looking to invade made more certain his prolonged imprisonment. Prevention of the certain invasion by the British of his own domain would have motivated the pragmatic soldier in Decaen far more than the dubious glory associated with publication of maps of a country Decaen knew France could never invade.

Also, despite a six-and-a-half-year head start over Flinders, Péron was having his own difficulties seeking that dubious glory.

CHAPTER TWENTY-THREE

THE RACE FOR PUBLICATION

Back in 1803 when Flinders had been approaching Mauritius's Port Louis, *Géographe* had only just departed for France. It took *Géographe* three months to reach the French port of Lorient on 24 March 1804. The *Naturaliste*'s earlier arrival had spread word in France about the difficulties with the expedition. Also one of the savants discharged at Mauritius on the outward voyage, Bory, had only recently published a book critical of Baudin. In the book he related the story of Bissy, the uncomprehending butt of Baudin's 'silver needle' quip, and used this as evidence of Baudin's lack of understanding of how compasses worked. The book had been very popular. Péron, still recovering from tuberculosis contracted in Cape Town during the return voyage, had a tough job in front of him. He had to justify in the eyes of the authorities the success of the expedition if he was to persuade them to finance the publication of a narrative and the charts of the voyage.

As Bligh had discovered, publication was impossible without the goodwill of the authorities; the account of his second breadfruit voyage still remained unpublished.

Péron and publication

In terms of currying favour, Péron was in his element. He brought into play the same skill by which he'd ingratiated himself with Jussieu and Cuvier, the gentry of Port Jackson, and which he'd applied to Decaen with less success. Péron immediately wrote to Decrès, the minister of Marine, advising that he was the expedition's sole surviving naturalist to return to France (true); that its collection of animals was unique (true, then and now; subsequently some became extinct); that most had been acquired by him (false); and all were under his care (true, but only since Baudin's death). His communication included the first of many backhanded slanders of Baudin and implied that Baudin had improperly removed animals from the collection at Mauritius – but Péron had now restored these. Péron then presented himself to Decrès, the first senior member of the expedition to do so. Its new commander, Milius, was still sick following their return. Péron also met with his old professor, Jussieu, still the powerful director of the Museum of Natural History. He leveraged this interest in the expedition's collection to generate official funding for the publication of an account of the voyage and its scientific results.

Péron was, as usual, indefatigable, chasing all avenues. He realised he needed support from the highest levels, and with Napoléon now enmeshed in continental war and planned invasion of England, Péron sought influence through his wife, the Empress Joséphine. She'd always had an interest in the expedition. On the eve of its departure in 1800, Baudin had been ordered to compile a collection of live animals for the empress, together with flowers, shells and precious stones, and in particular birds with beautiful plumage.[1] The empress's bird keeper escorted Péron's caged animals when they were moved from the coast to Paris, and a large quantity of these animals were housed at her estate at Malmaison, outside Paris, as were all the expedition's native artefacts. The artefacts included those donated to the French by George Bass in Port Jackson specifically for a proposed museum the Society of the Observers of Man, which was never built.

As 1805 wore on, Péron's need for personal finances was becoming great. He was writing his own unofficial account of the voyage, prompted largely by some verbal encouragement from Decrès, although no promise of finance had been made. Baudin's own journals were regarded as irrelevant. Bory and Péron had sufficiently debased Baudin publicly, and the authorities had no interest in anything he'd written. The few members of the expedition to support Baudin were either already dead (Maugé, Riedlé, Levillain), marginalised (Ronsard) or too junior to matter (young midshipman Charles Baudin, now a lieutenant on a French frigate). None had power to resist the unwavering force of Péron's campaign against Baudin's character, and they had their own careers to consider. History is only slowly correcting this.

By mid-1806 Péron was desperate for income and persuaded the Institut National to uphold his claims for financial support. It's no mystery that the names of the signatories to the letter – Fleurieu, Lacépède, Jussieu and Bougainville – would figure prominently in place names in the accompanying atlas when it was published. The report itself was replete with Péron's hyperbole, claiming that the expedition had discovered 2500 new zoological species whereas Cook, its most brilliant precursor, had produced no more than 250, and that Péron and the artist Lesueur had been exclusively responsible for these. As Péron had already, unofficially, written much of the narrative for the voyage, it isn't surprising that – no doubt through Péron's influence with Jussieu – the National Museum recommended immediate publication of the narrative of the voyage, its atlas and its scientific discoveries.

Baudin's name damned

Péron also sought support from the Ministry of the Interior for publication of the various results of the voyage. It seems the Ministry of Marine and the navy wanted little to do with the expedition, other than the publication of its charts. It's unclear whether this had to do with a desire to distance themselves from Baudin or Péron or to their distraction with the war and the current publication of d'Entrecasteaux's voyage following the return from exile in London of Rossel, its surviving commander. In any case Péron finally got his precious funding, both to support Lesueur and himself while he wrote the narrative and to pay for Freycinet to prepare and have engraved the various charts of the voyage. This was granted by decrees from Napoléon, on 4 August and in October 1806, more than two years after his return to France. The decrees made no mention at all of Baudin.

His name was officially damned. Napoléon is rumoured to have said, 'Baudin did well to die; on his return I would have had him hanged.'[2]

Terre Napoléon and the atlas with no charts

So began the French publication that ultimately produced the first chart ever of Australia with its coastlines largely complete.

The publication of the narrative of the voyage was to be in three parts, covering first history, then anthropological science, physical science and meteorology, and finally navigation and geography. Natural history would only be published through public subscription – an unlikely event given the troubled times.

Ex–gunner's mate 4th class and the expedition's natural history painter, Lesueur, would be responsible for most of the artistic publication, although he'd painted predominantly flora and fauna. Petit, the other ex–gunner's mate 4th class and the expedition's portrait/zoological painter, who'd produced the striking portraits of the Aborigines of Van Diemen's Land, would never complete the full publication of his images of native people. He died of gangrene after a street accident in 1804, two weeks away from his intended marriage. Louis de Freycinet agreed to help with the coastal charts, as illness prevented him from further service at sea for the time being. Their drawn-out production and completion, followed by the expensive process of engraving and preparing the charts for printing, would consume all available financial and energetic reserves of Péron, Lesueur and Freycinet.

In 1807 Péron published the first volume of the historical account, *Voyage de Découvertes aux Terres Australes* – he was still writing the next volume – together with the first part of a larger accompanying pictorial atlas, entitled *Atlas Historique*. It covered the voyage only as far as Port Jackson. Baudin's name didn't appear in it at all; he was referred to only as 'Commandant'. Flinders, detained in Mauritius, would read of the publication with frustration, envy and not a little anger. Péron's *Voyage* claimed for France the stretch of coastline discovered and first examined by Flinders from the top of the Great Australian Bight, at the islands of St Peter and St Francis, to Encounter Bay at the mouth of the Murray River and the Victorian coastline first charted by Grant, as well as the smaller portion discovered by Baudin.

Péron had named this entire stretch of coastline Terre Napoléon, and the *Voyage* gave French names to all the features west of Port Phillip and elsewhere. An intense Napoleonic strain was evident, in particular the names of all members of the regime who'd supported Péron and publication. These included Napoléon; his wife; his brother; his famous victories; Decrès, the minister for Marine; the Institut National, after which an entire archipelago was (and still is) named; as well as its more famous members, Bougainville, Lacépède, Fleurieu and Jussieu. The name 'Péron' also appears, several times. Even Thevenard, the prefect of the port of Lorient who'd allowed Péron leave to visit Decrès so soon after his return to France, wasn't forgotten. Baudin's name was mentioned nowhere at all.

Flinders could only fume. There he was, still imprisoned on Mauritius. He hadn't yet completed, let alone published, anything from his expedition other than preliminary charts – now languishing in a drawer in the Hydrographer's Office in London – and certainly had given no major names to much of the coastline he'd surveyed. Admittedly Péron had little alternative but to insert his own names on the charts as Flinders hadn't published more than two or three names to any part of the coastline. However, there was no acknowledgment anywhere of Flinders' earlier discovery of these coastlines, of which the French were only too well aware.

Although Péron's *Voyage* claimed most of Flinders' exploits along the south coast for France, neither the *Atlas Historique*, despite its contents page, nor the *Voyage* contained any charts of these coastlines, nor for that matter any others. These were still being drawn, and in 1807 were less developed than Flinders' were at the same time. Instead the *Atlas Historique* contained engravings of many of Lesueur's and Petit's paintings and a small plan of Port Jackson. Péron had successfully pre-empted Flinders' right to name the parts of the coast, a right Flinders claimed as first discoverer. He'd done so without even producing a complete chart of the coastline. To make it worse for Flinders, this coastline now boasted the names of his enemies – those responsible for his own imprisonment.

As one further indignity for Flinders, Péron's narrative debunked the theory of the North–South Channel across Australia, a theory that Flinders had been the only person to completely disprove, having explored both its southern and northern coastlines. No Frenchman had visited the Gulf of Carpentaria at all. Flinders could no longer be the first publisher of this either.

CHAPTER TWENTY-FOUR

THE FRENCH PUBLISH THE FIRST MAP OF AUSTRALIA'S COASTLINE

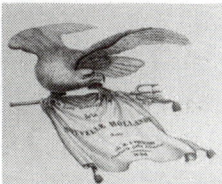

A chart of the complete coastline of Australia was yet to be published. Baudin and Flinders could both claim to have filled in gaps in the coastline left undiscovered by past navigators, but no-one had yet produced a definitive chart to prove it. This historical event occurred in 1811 when the second part of Péron's *Atlas Historique*, in the form of a smaller quarto atlas, Freycinet's *Atlas, Deuxième Partie*, was published.

The spur for the much-needed injection of government funds that allowed Freycinet to complete the charts and for their consequent publication was probably Flinders' release from Mauritius in 1810. The desire of the French to get their charts out before Flinders could publish his is evidenced by a letter from Freycinet in August 1811 to Decrès, the minister of Marine, urging publication:

If the English publish before the French the records of discoveries made in New Holland, they will, by the fact of that priority of publication, take from

The first published
complete map of
Australia, from the
*Atlas, Deuxième
Partie*, 1811

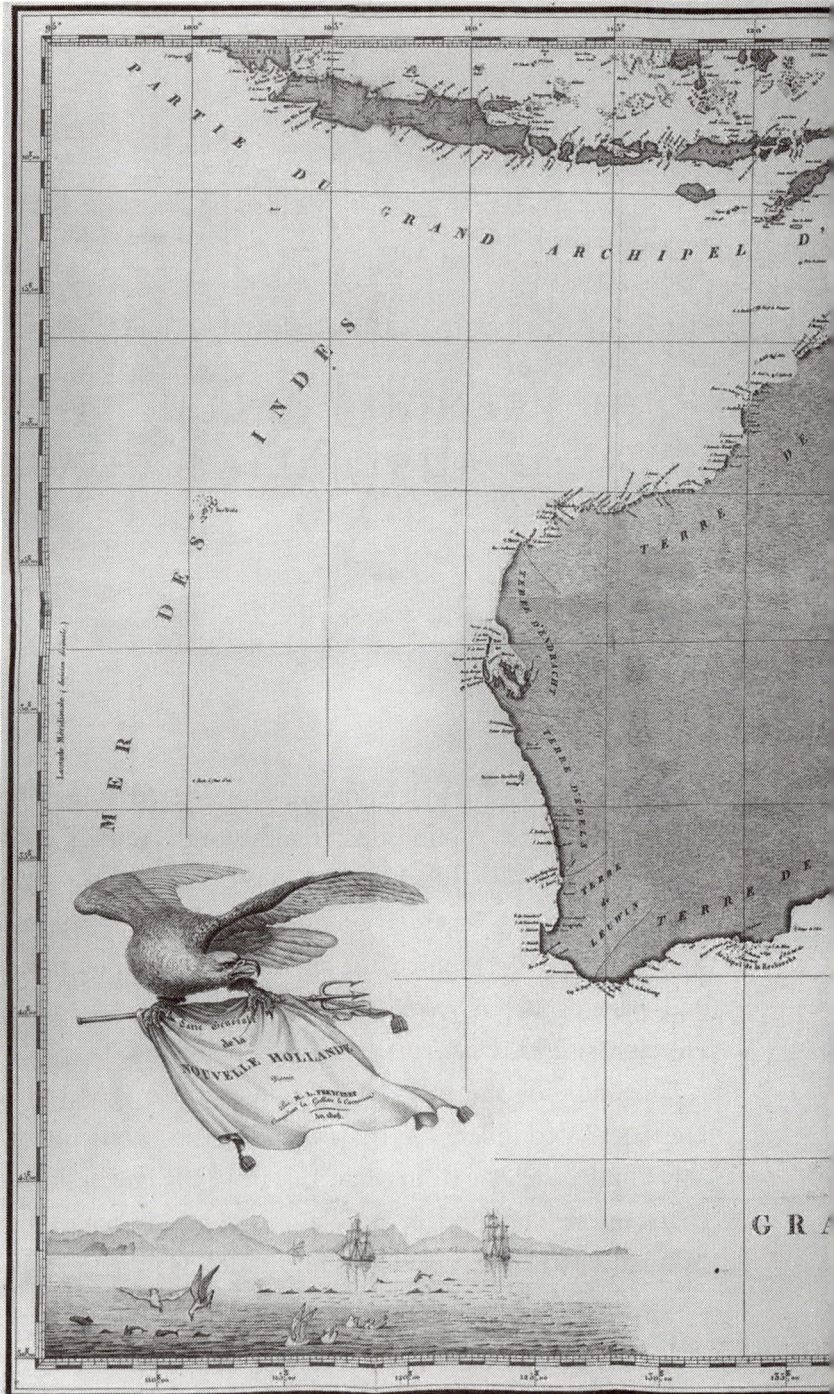

Cape Decaen (on Fleurieu Peninsula) and Decaen Island from Plate 7, Freycinet's *Atlas, Deuxième Partie*, where Flinders recorded Encounter Bay

us the glory which we have a right to claim. The reputation of our expedition depends wholly upon the success of our geographical work, and the more nearly our operations and those of the English approach perfection, and the more nearly our charts resemble each other, the more likelihood there is of our being accused of plagiarism, or at all events of giving rise to the thought that the English charts were necessary to aid us in constructing ours; because there will be no other apparent motive for the delay of our publication.[1]

In the *Atlas, Deuxième Partie*, Freycinet's charts were beautifully executed and adorned with sketches of kangaroos, dwarf emus and eucalypts, with the expedition's ships in the background. The *Atlas, Deuxième Partie* contained plans and charts of Terre Napoléon, including Golfe Bonaparte and Golfe Joséphine, Péron's names for Spencer Gulf and the Gulf of St Vincent. It has amused some historians that Péron named after the empress a gulf that Baudin had initially named Gulf of Misanthropy on his chart.[2]

Péron had given some other land features names that contained nasty little stings specifically aimed at Flinders.

Cape Decaen and Decaen Island, a dagger to Flinders

On plate number seven of the *Atlas, Deuxième Partie* Freycinet had named Cape Decaen, flanked by Decaen Island, commemorating Flinders' hated captain general, the man who'd imprisoned him for six-and-a-half years.

Cape Decaen and Decaen Island flank the expanse of water that Flinders had called Encounter Bay on his charts, the historic place where the *Investigator* and *Géographe* had met so many years before. Decaen's name appears nowhere else on the coastlines in the *Atlas, Deuxième Partie*.

Was this some idea of a joke on the part of Freycinet and Péron, poking fun at the rival they'd met at Encounter Bay who'd been incarcerated for so long by Decaen, who now shared his name with this flanking cape and isle?

Flinders couldn't have helped notice this, as Encounter Bay would be the first place he would have examined on the French chart for some recognition of his own efforts on that coast. It would have delineated the border between his discoveries and Baudin's. Only from Encounter Bay south should any Terre Napoléon have, to his mind, existed. There was nothing on the French chart in that region to recognise Flinders' work other than his association with the name of Decaen. This would have cut Flinders deeply.

To publish this was a particularly mean act of Freycinet's. It's not likely to have

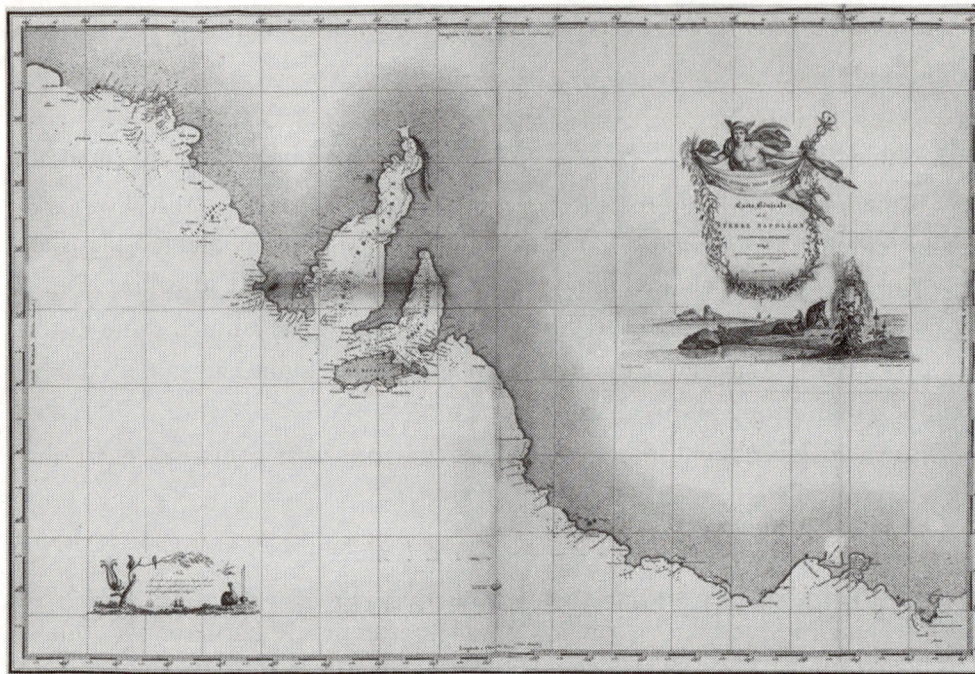

The chart in *Atlas, Dexième Partie* that so offended the British,
proclaiming Terre Napoléon

been a simple oversight, given the preoccupation with names (for example, exclud-
ing Baudin's name from the publications entirely). No wonder Flinders believed that
it was the French Government influencing Péron and others, rather than those individuals themselves who were responsible for the injustices of Péron's publications.

By now there was a waning of impetus for publishing the outstanding French
historical and scientific volumes and remaining full-scale charts; only reduced sizes
were published in *Atlas, Deuxième Partie*. Péron's health had never fully improved,
and tuberculosis, initially contracted in Cape Town on his return voyage from
Mauritius, had resurfaced.

Péron died on 14 December 1810 with the second volume of his historical narrative unfinished. The balance remained incomplete, although Péron had done
much. Freycinet was ultimately appointed to this task in 1815. Lesueur, co-author
appointed under the decrees of 1806, left for America, where he was to enjoy a
much more successful artistic career.

It's a final irony that the *Atlas, Deuxième Partie*, proclaiming boldly Terre Napoléon and liberally littered with Bonapartist place names, would never ingratiate Péron and its other authors with Napoléon as intended. A comprehensive atlas of forty-four detailed navigational charts was produced by Freycinet in 1812, but within a matter of months afterwards Napoléon's Grand Army had littered its frostbitten entrails on the retreat from Moscow in the disastrous winter of that year. Napoléon would shortly suffer defeat at the Battle of Leipzig in 1813, abdicate and head to exile on Elba in 1814. There was no longer anyone in power to cultivate, and it wasn't the time to be known as a Bonapartist.

Why the lack of interest in the expedition in France? Even by the time it'd returned in 1804 it was no longer relevant. The mood in France had changed from those exciting closing years of the previous century. The expedition had left France in the hands of a successful Napoléon, flush with victory and with more to come. Invasion of India and beyond – including perhaps Port Jackson – and England, were probabilities despite the French naval disaster at the Battle of the Nile in 1798. The Treaty of Amiens in 1802 gave Napoléon breathing space to rebuild a better base for attack. Péron and the expedition returned in 1804 to a France under British naval blockade, with a French fleet that was still not rebuilt and would be almost totally destroyed by Nelson at Trafalgar in October 1805, obviating any possibility of overseas campaigns, whether in England, India or Port Jackson.

Against this backdrop it's easy to see why Péron had to labour as he did for publication. It's a tribute to his tenacity that he succeeded at all. He did so at the cost of his health, dying at the young age of thirty-five while living in a cowshed in his home town – as was the tradition for invalids – still short of funds and still chasing the fame that the full publication was to have provided.

CHAPTER TWENTY-FIVE

FLINDERS PUBLISHED AT LAST

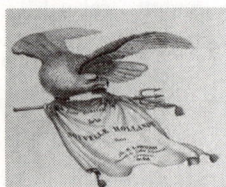

As the *Atlas, Deuxième Partie* was selling in Paris, Flinders was treading the same wearying path to publication in London. He'd arrived back in London in 1810 a totally different man physically. His hair, once dark, was now white. Although only thirty-six, he looked fifty. Young John Franklin, his midshipman from *Investigator* and future tragic polar explorer, had briefly witnessed Flinders' reunion with his wife, Ann, before feeling so overcome that he retired to leave them alone.

Captain Flinders, a disappointing promotion

So many of Flinders' career aspirations were not so much dashed as only half-fulfilled. Before his departure in *Investigator* back in 1801, Lord Spencer had promised him promotion upon his return. Flinders was promoted to the important rank of post-captain only as of 7 May 1810, the commencement date of the term of the current First Lord of the Admiralty. Flinders wanted this backdated to 1804, the date of his expected return to England, but for Decaen. His efforts included considering an appeal to the House of Commons and an order for back-dating the promotion from the king-in-council. This ultimately alienated many supporters in the Admiralty and, for a time, Banks. They were also unsuccessful; this remained a cause of bitterness until his death.

The Admiralty's actions were based purely on precedent; no-one was promoted during imprisonment. They weren't impressed by Flinders' argument that technically he wasn't a prisoner-of-war. In the midst of the long and expensive war with France, with its frequent naval action, there were many with Flinders' complaint, and it was in the Admiralty's interest to stick to the precedent. Post-captains cost more than junior officers, and to promote those junior officers and pay them while in prison and not actively serving His Majesty wasn't affordable. The result was that there were many younger men who had seniority to Flinders on the post-captains' list. They would be preferred for postings; they would receive flag rank before Flinders; and if the war ended, there would be a large number of post-captains from the wartime navy eking out an existence ashore on half-pay. The lower down the list one was, the worse were one's prospects for advancement.

Flinders didn't endure this alone. Fowler, Aken, Franklin and Samuel Flinders had little to look forward to as they could only be promoted on the basis of their conduct during the voyage. This in turn was only possible once Flinders had made his full report on their conduct, which he hadn't completed during his years at Mauritius. Their careers also remained on hold.

Flinders' physical well-being was also compromised. Scurvy had initially been the cause of the deterioration in his health at Mauritius, but then his 'gravelly complaint' returned. This was a kidney or bladder infection that had initially appeared in 1795 and had recurred probably as early as 1804, with painful consequences. Carrying these various strains he began the steps to publication of his narrative and charts.

Publication but Admiralty parsimony

Unlike Péron, Flinders had little difficulty finding political support for publication, although his personal finances were a different story. With its interests in Port Jackson and, by now, Van Diemen's Land well established, the Admiralty was keen to publish his charts, notwithstanding the earlier publication by Freycinet of the *Atlas, Deuxième Partie*.

The Admiralty did, however, quibble over his pay during this time – it was only half-pay of 4 shillings per day – as well as over such trivia as the minuscule cost of sugar and wine used to treat the sick on *Investigator*, for which Flinders was technically accountable, and her missing purser's books, lost in the sea at Wreck Reef. The East India Company was more commercial, paying without demur the £600 table money it'd promised in 1801; this Flinders distributed among his officers, Brown, Bauer and Westall, leaving him £300 as a necessary income source. Luckily there was still Banks's support. Other benefactors appeared as well. For example Bligh, now an ageing rear-admiral, introduced Flinders to the Duke of Clarence, brother to the reigning and profligate Prince Regent, in power following the onset of King George III's madness.

On 16 January 1811 the Admiralty approved the writing of the narrative and preparation of an atlas of charts under the supervision of Flinders. However, with typical single-mindedness, it was only the atlas, containing its valuable sixteen charts, for which the Admiralty undertook to pay. Flinders was told that he would only recoup his costs for writing the narrative from subscription payments – essentially he had to take publication risk.

This was a considerable contrast to the French experience. It may be that Péron wasted much time and his health in obtaining funding, and that the French Ministry of Marine, like the Admiralty, was interested only in Freycinet's charts, but thanks to the support of the French Ministry of the Interior it was only Péron's chapters on natural history that had to be paid for by public subscription. Bankrolling the publication of just the charts was a little ungrateful of the Admiralty, seeing that so much of Flinders' descriptive advice and sailing directions, for example on sailing through the Great Barrier Reef, on 'threading the needle', would be contained in the narrative. So although Flinders had some financial support, it was much less than Péron could obtain, although obtained so much quicker.

The incorrect almanac and more delays

Flinders set about readying his material for publication. This took three years. His diligence meant much financial distress was incurred by the post-captain on half-pay.

To complete his charts Flinders, or more specifically his brother Samuel, had made thousands of lunar observations at sea to verify the longitudes that Flinders had calculated through sextant observations and chronometer readings. Flinders had been very precise, using both lunar readings and chronometer calculations to calculate his longitudes because of the debate still raging at the time as to which was the more correct way to calculate longitude.

The lunar school believed that longitude could only be established accurately by reference to the moon's observed position east or west of a star at, say, noon, when compared to predicted positions of the moon and the star at Greenwich. The Royal Observatory at Greenwich had calculated massively complex tables predicting what those positions at Greenwich would be for these purposes for some years forward. These were published in the *Nautical Almanac*, and seamen relied on them. The lunar school didn't believe that a clock could be developed with sufficient accuracy to record Greenwich time reliably for use as a comparison with local time in order to determine longitude. There was some truth in this, which was why the navigators were constantly taking accurate astronomical observations to check whether their chronometers were gaining or losing time and, if so, whether at the same or a different rate.

The lunar tables were extraordinarily complex, requiring a keen mathematical mind to accurately work them. Cook didn't have a chronometer for his first voyage. He relied on 'lunars' as much as he could, but more often on dead reckoning. Baudin constantly had problems with his chronometers and relied a great deal on astronomer Bernier for 'lunars'. Although Flinders did have a chronometer – in fact he had three – it's a testament to his thoroughness that he also used lunar observations, presumably performed by Samuel in the course of his astronomer's duties, and the *Nautical Almanac* almost as much.

In a stroke of misfortune, the *Nautical Almanac* was found to contain significant errors but this was realised only after Flinders' departure in *Investigator*. Flinders' thousands of calculated positions, crucial for his charts, had been based on the almanac so were therefore also wrong. They all had to be recalculated from the original observations jotted down in the expedition's daily records. Brother Samuel, now on half-pay and with his naval career recently ruined by a poorly handled court martial, was retained to perform this massive task under the supervision of former astronomer to the voyage, John Crosley. Crosley had left the expedition ill at the Cape on the outward-bound voyage, and Samuel had taken over his duties.

Flinders' discovery, magnetic distortion

There was one other matter that delayed compilation of all the bearings required for completion of the charts. Flinders wanted to ensure that his calculations weren't distorted by a phenomenon that no-one else had ever noticed or had allowed for. This was the effect of ships' magnetism on all compass observations.

As noted earlier, during his detention on Mauritius Flinders had written several papers on this subject that had been published by the Royal Society. In early 1812 the Admiralty permitted him to conduct experiments which proved that the various sources of iron on a ship would seriously distort the compass readings taken. He suggested a means of rectification – an adjustable soft iron bar inserted vertically below the compass on the binnacle could counteract the effect of the ship's iron around it. This became known as the 'Flinders Bar' and was integral to a ship's navigational gear for the next 150 years. It was a significant contribution to maritime navigation and, while important for the age of wooden ships, its impact would be even greater, although Flinders couldn't know this, once the age of iron-hulled ships commenced after his death.

These delays meant that the first finalised charts weren't delivered to engraver Aaron Arrowsmith until 1813, and the last of the charts to be engraved weren't completed until June 1814. The atlas was to be printed in large and small sizes and its sixteen meticulously calculated charts would be far more accurate than those of Freycinet's *Atlas, Deuxième Partie* or the subsequent navigational charts Freycinet produced from these in 1812. In addition it included two plates, depicting Westall's twenty-eight coastal profiles, and ten plates of Bauer's botanical drawings.

Flinders was also writing his official narrative. By October 1812 the first of two volumes of the narrative was almost ready for the printers and the second had been roughed out. Nine of Westall's works would be included.

Australia and Australians

Flinders was the originator of Australia as a place name for what had always been called New Holland. It comprised an amalgamation of several descriptions used in the past for suspected landmasses in the region over the preceding 200 years. At one stage he proposed Australia as an alternative to Terra Australis in the title to his narrative, *A Voyage to Terra Australis*. Because Flinders had proved New Holland and New South Wales – as the western and eastern areas of Australia were then named – were the same landmass, with no dividing north/south channel, it was important to him that the landmass have only one name. He referred to it generally in his narrative as Terra Australis, but only because Banks and others made him. He says in a footnote:

> Had I permitted myself any innovation upon the original term, it would have been to convert it into AUSTRALIA; as being more agreeable to the ear, and an assimilation to the names of the other great portions of the earth.[1]

Despite the objections of others, as we've seen Flinders did manage to refer in his narrative to Aboriginal people as 'Australians', of itself an important step.

The huge amount of work in writing and co-ordinating production of his narrative and the atlas further taxed Flinders' delicate health. Inadequate income for a man who'd always been keen to ensure he was sufficiently financially resourced, together with the birth on 1 April 1812 of his only child, Anne, would've added to the pressures.

The return of the 'gravelly'

This wasn't the worst of it. In late 1813 Flinders' 'gravelly complaint' – the kidney or bladder infection – returned with a vengeance. This, the third occurrence, was serious and still totally undiagnosed. By early March 1814 he could no longer work; by late March the surgeon was visiting every day. On 19 July 1814 his devoted wife, Ann, placed the two newly bound and published volumes of Flinders' narrative on his bed, but he never opened them. He died that day, aged forty years four months.

Whatever did kill Flinders, it was neither Decaen nor the work on the narrative. Like Péron, Baudin and other navigators, the illness was a legacy of his travels. This could well go back to his days with Bligh in *Providence* and, according to Ingleton, mercury-based treatment administered to him for 'venereals' – probably gonorrhoea. There may have been recurrences of this condition during his two years onshore at Port Jackson when with *Reliance*, resulting in urethral stricture and chronic urinary retention. Alternatively the cause may have been entirely different and resulted from six-and-a-half years' imprisonment on Mauritius. Péron wrote in his *Voyage* that inhabitants of Mauritius were generally affected by:

> … all distempers of the urinary passages … to an extraordinary degree; they seem to proceed from the quality of the water which … contains a great proportion of carbonate of lime.[2]

The cause may have been Flinders himself, and a wrong diagnosis. Flinders' background as a surgeon's son may have caused him to attempt his own early diagnosis, erroneously believing, according to Samuel Flinders,[3] his illness to be a stone in the bladder. Flinders, one could speculate, was his own worst enemy, not only in relations with others, such as Decaen, but also in assessing his own physiology.

In a final blow, the narrative didn't sell well. Napoléon had abdicated only months before, the occupation of France was in full swing, and Australia attracted little interest. The book was remaindered with less than half the print-run sold. Ann Flinders had to pay out the printer's bills – the Admiralty took no publication risk – and led an impecunious life for the next thirty-eight years until her death in 1852.

CHAPTER TWENTY-SIX

THE ACHIEVEMENTS

Flinders and his colleagues

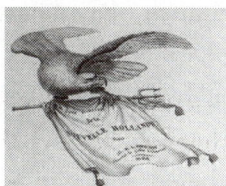

Flinders' labours produced an atlas with sixteen charts so accurate that they remained in common usage, in some cases, for over a century. This was even despite the fact that Flinders hadn't charted much of the north-west Australian coastline and largely relied on the earlier Dutch charts and Freycinet's publication to create these.

Botanist Robert Brown had collected over 3900 plant species, of which at least 1700 were unknown to science. He complained that he hadn't been given sufficient opportunity to collect as he wished, though one wonders if his assistant, Peter Good, would have agreed. Despite Brown's despondency, his toil in carefully noting and preparing his relatively meagre findings for publication would far outweigh Péron's results.

Leschenault, the surviving botanist from the Baudin expedition, had remained at Mauritius until 1807. Péron had therefore been the sole custodian in France of the expedition's scientific collections and records. Péron jealously guarded the records and much of the collection, there being no-one else with any competence surviving in France for whom they had any relevance.

Following Péron's death, expedition artist Lesueur took many of the records with him to America, the upshot being that the zoological notebooks of Lesueur's drawings weren't available to other scientists until well after 1874. The botanical results were ignored. In Leschenault's absence, much of the remaining botanical work was taken over by Labillardière, surviving botanist of the d'Entrecasteaux expedition, who proceeded to act as if it were his own personal property. Consequently, on his death the expedition's collection was all treated as part of his estate and sold … to an Englishman.

In contrast, Brown, now employed as Banks's personal assistant and librarian in London, completed thorough documentation of over 3600 plants from the *Investigator*'s expedition and, in 1810, accompanied by Bauer's paintings, was able to publish descriptions of over 2000 species, 1500 of these new to science. This was despite his complaints that he'd had insufficient opportunities ashore. Again, the stability provided by Banks's resources ensured a useful result for Brown. Had they been completed and published, the French collections would have easily rivalled the English, but this was never to be.

Artist Ferdinand Bauer produced 1000 drawings of plants and 200 of animals from the *Investigator* expedition, all of use to botanists, unlike much of Lesueur and Petit's work, despite their artistic beauty. Bauer remained in New South Wales after Flinders' departure on *Porpoise* and produced a further 1070 drawings. Fifteen plates were published in 1813, but Bauer's standards for colour were beyond any printer of the time so he set about to handcolour the edition, which didn't continue. Bauer's productions, being handcoloured, are extremely rare and valuable. A copy of his *Illustrationes* held by the British Museum was recently valued at over £4 million.

Finally, Flinders left a successor, in the tradition of the Cook's school, in young midshipman John Franklin. Franklin would later feature prominently in Australian history as the governor of Van Diemen's Land, and also in polar exploration. In 1848, by then Sir John Franklin, he disappeared with his ships *Erebus* and *Terror* and all his 128 men, leading an expedition searching for the north-west passage that had eluded Cook sixty years before.

Baudin

Freycinet succeeded in publishing the first complete map of Australia and the first charts of the South Australian coastline in 1811 following Péron's death, with accurate nautical charts to follow in 1812. Flinders' more detailed and accurate maps wouldn't be published for some years. However, as we've seen there was no longer much interest in France in the expedition.

Large chunks of the extensive collections from the expedition were either overlooked or wasted – in the grounds of Malmaison or in the hands of Péron during his illness and Labillardière following his death. The same could be said for other activities of Péron. His oceanographic records were probably the first concrete efforts at consistent scientific measurement of the time. Péron's rudimentary deep-water measurements provoked only ironic comment from Baudin but have interested scientists more recently, particularly in relation to global-warming issues.

The expedition's achievements in anthropology, particularly from Van Diemen's Land, are without equal, but again publication was delayed or confounded by the absence of most of its drawings until released by Lesueur's estate in the late nineteenth century. Nevertheless, Baudin's expedition had far more contact with the Tasmanian Aboriginal people than any other expedition. The tragic extinction of that race since makes records of these meetings unique.

Péron's study of marine invertebrates broke entirely new ground. Lesueur and Péron had planned a separate publication on the subject, again confounded by Péron's death. This important field of study would sadly remain relatively ignored for a further half century.

Finally, in Louis de Freycinet the French had a budding navigator whose services would provide a base for Pacific exploration over the next thirty years.

CHAPTER TWENTY-SEVEN

CONCLUSIONS

The legacy of the navigators, and the future

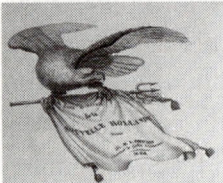

The Treaty of Paris, the conclusion of the American War of Independence and the French Revolution caused a massive shift in the opening up and exploration of the Pacific in the last half of the eighteenth century, culminating in the settlement of New South Wales and the expeditions of Flinders and Baudin. They and the earlier navigators carried out the initial groundbreaking explorative work at massive personal cost. Next would follow the more painstaking and sometimes more dangerous task of intricate survey.

The conclusion of the Napoleonic Wars at Waterloo in 1815, in conjunction with the commencement of rational organisation of the British Naval Hydrographical Office after the departure of old Dalrymple, would precipitate this more methodical and exact survey work for the first half of the nineteenth century. Waterloo didn't influence this directly. Its immediate effect was to release from service hundreds of thousands of sailors; their officers were left onshore on half-pay. In this light hydrography and surveying were suddenly appealing ways to maintain full pay, obtain a command and work towards promotion in peacetime. It was for similar reasons that Cook had become so well qualified for the *Endeavour* voyage following the Seven Years' War.

The riddle of the Australian rivers had still not been solved. The work of Flinders and Baudin, having not produced any large rivers, only compounded the mystery. Many explorers still harboured suspicions of an inland sea or that large parts of the northern coastline might be separated from the southern mainland. These would, if true, provide chances for new settlement and exploration, and possibly a lucrative support base for Indian, East Indian and Chinese trade. It was important to the British to find out exactly what was there before any other interested nation. It was important to the French, following Waterloo, to seek out new sources of trade, as it had been after the Treaty of Paris.

In 1817 Phillip Parker King would seize the opportunity to chart the difficult north-western Australian coastline for the British as an alternative to life ashore on a lieutenant's half-pay. His father had entertained Baudin in 1802 when governor of New South Wales. Louis de Freycinet, responsible for the charts in the *Atlas, Deuxième Partie*, would be sent by the French in 1817 to further survey the west Australian coastline. Again, the careers of two officers would be determined by national rivalry and a need for expanded trade, precipitated by the end of these wars.

Who were the winners and who were the losers in all of this? The losers are pretty easy to spot. They are the navigators.

The French

Almost none of the French navigators survived. Bougainville and Louis de Freycinet were the only ones. The conflicts of the late eighteenth century certainly encouraged French exploration but chewed up the executives responsible for it.

Surville and Marion had been, respectively, drowned and eaten.

Kerguelen was jailed in disgrace, and St Allouarn was dead.

La Pérouse and his entire expedition had been lost without trace.

D'Entrecasteaux and Kermadec had perished and their expeditions had disbanded and scattered.

Baudin was dead, his expedition disbanded and irrelevant to France.

The French winners were their future surveyors who followed Baudin. The efforts of Baudin's expedition in 1803 would lay for France the same foundations that Cook had set down for Britain in 1779, in the form of a body of experienced surveyors.

Baudin's subordinate, Louis de Freycinet, would lead his own expedition in 1817. Interestingly, for such a close colleague of Péron and a critic of Baudin, he refused to have civilians on board his own expedition. In the tradition of French exploration he also smuggled his wife, Rose, on board with him, against regulations, much as Flinders had planned to do on *Investigator*. Freycinet retired as an admiral.

Freycinet's second lieutenant, Duperrey, would sail to the Pacific in 1822 in command of his own expedition. Another of Baudin's better-connected subordinates, Hyacinthe de Bougainville, would lead an expedition in 1824.

The culmination of these would be an expedition to the Pacific in 1826, led by one of Duperrey's lieutenants, Dumont D'Urville. It surveyed much of the Pacific, including the final resting place of La Pérouse's ships at Vanikoro. His work led to French colonisation of a number of those island groups.

Freycinet's voyage would cause the British to send Phillip Parker King on a north coast survey, thus encouraging a lineage of future British surveyors.

The British

Were the earlier British navigators in any sense winners? Although surviving a little longer than their French counterparts, they really didn't fare much better.

Cook died at a stage in his career at which he was due for retirement anyway. Nevertheless he died well before the man to benefit from his labours, Banks. Cook's significant contribution for this story was 'Cook's school' of young navigators, particularly Bligh and Vancouver.

Vancouver, although a brilliant and diligent hydrographer with interesting global political foresight, died a young, sick man, although his career was already largely destroyed by Baron Camelford and his domestic political factions.

Bligh was the great survivor. Politically exiled following the *Bounty* mutiny and despite his own demonstrated courage and seamanship in two famous sea battles – he earnt a commendation from Nelson himself – and two further mutinies,

Bligh was still asking Flinders to dedicate his narrative to Bligh late in life. Flinders did the astute political thing, ignored Bligh and dedicated the work to all the First Lords of the Admiralty in office during *Investigator*'s voyage. Bligh died a rear-admiral, with a large family, in quiet retirement, but still earlier than the main beneficiary of his labours, Banks.

Flinders was a man with a terrific mind and surveying skill. However, he was a slave to his own stubbornness, pride and obstinacy and suffered for it. He was, nevertheless, a zealous and tenacious officer who did the utmost for his country.

Grant and Murray were consigned to history's archives and both probably died in quiet anonymity on a lieutenant's meagre half-pay. Thistle and Taylor, casualties of Cape Catastrophe, never had a chance to shine, and Samuel Flinders resigned his commission after Matthew's death.

The single man to benefit from the navigators' work was Banks. Of course, without Banks the British navigators would've achieved little. For over forty years he was their patron when he needed them – and usually when the positions were reversed. Although Banks could be loyal when needed, as he was with Bligh and Flinders, he was the nemesis of those who crossed the line and spurned him or those in his circle, as had Vancouver with Menzies. He was all-powerful during their life span. And he was the greatest direct beneficiary of their work. They entrenched his position as unofficial minister for science to any British government, and as president of the Royal Society.

The greater indirect beneficiary, in Banks's mind, was the State itself. Through Banks, and his beliefs in Enlightenment serving the State, it was Britain that obtained the full benefit of the navigators' work. And maybe Banks's beliefs in 'Enlightenment' in service of the State were right. Not all the results were successes, of course, but this must be expected where new frontiers are involved. Cook never found the North-West Passage, but discovered that the fabled *Terra Australis Incognita* was nothing more than a frozen Antarctic continent. Bligh's breadfruit experiments were ultimately failures. Flinders proved that there was no north–south passage through continental Australia, destroying hopes of a new trade route to India, and was unable to survey a practicable passage through Torres Strait. Baudin, while doing more survey work on the north-west coastline than any other navigator to date, didn't produce any tangible material benefit from it.

As a result of the British navigators' labours, and the political upheavals that fostered them, the colony of New South Wales was born. Through Banks's supervision – and the labours of the British navigators – that colony, and others that now make up the Australian nation, flourished. Coastlines were delineated; ultimately navigable sea lanes were laid down to support its initial survival and its subsequent development and expansion. Most of the British navigators, being navy men in service of the State, would have applauded these results.

There was also general benefit from publication of the scientific and geographic results of all these expeditions. It's interesting that the British and French, although enemies for most of the navigators' lifetimes, were still prepared to share their results in this way and issue passports of free passage to allow these expeditionists to proceed in times of war. This tradition of scientific maritime survey, now entrenched in Admiralty practice, would remain significant for most of the nineteenth century.

On a broader level they were also expanding knowledge of their own environment and how to survive in it. These voyages carried the first specialised scientists to consider these broad topics, whether as mineralogists, botanists, zoologists or, as in Peron's case, in more esoteric roles such as comparative anatomists or anthropologists. Many more would follow.

Banks died in 1820, well after all the navigators. By this time, a lineage for development of coastlines – not only of Australia's but also others – had been established directly from their labours. John Franklin, Flinders' cousin and midshipman, would conduct his own polar exploration and also influence Phillip Parker King. King, the first Australian-born British rear-admiral, in turn would effectively support a renewed surveying 'school', influencing surveyors such as FitzRoy, and through FitzRoy, Wickham, Stokes, Stanley, Denham and others. These men would complete much of the unfinished surveys of the northern Australian coastlines commenced by Flinders and Baudin. They would go on to chart other coastlines in the Pacific and along South America. Their successors took the next step, for example charting the sea bottom itself during *Challenger*'s voyage in the 1870s, effectively carrying on those early experiments of Johann Forster on *Resolution*, and Péron on *Géographe*. The early efforts of the navigators laid essential foundations for this.

But the indirect consequences of these efforts would go further than hydrography or even Enlightenment in the service of the State and create an outcome that's still being digested now. The naturalist from one of FitzRoy's voyages would ultimately publish his views of the world and its development following three years with this able surveyor and his ship. The ship was the *Beagle*, the naturalist was Charles Darwin, and the result was entitled *Origin of the Species*.

$\mathscr{A}$PPENDICES

Appendix I: A note on names of people, ships, places and distances

French names have been abbreviated to what appears to be general practice or applicable for particular individuals.

Ships' names have been in all cases expressed without any title, such as HMS. Given the unusual nature of some ships and their proper names (e.g. HM Bark *Endeavour*, and HMAV *Bounty*), accuracy has given way to pragmatism.

Names for places have changed over time, particularly with 'Terre Napoléon'. If a place name changed to a modern equivalent during the period discussed, then I have tended to use the modern term. The obvious example is 'Australia' instead of 'New Holland', but Tasmania has remained Van Diemen's Land, as that is what it was still known as at the time of Banks's death in 1820. To assist with locations I have often referred to the current name in brackets.

Distances have been given in metric except where either inappropriate (if a small amount), concerned with the length of a ship or part of an original quotation.

Other terms I have tried to sum up in a short (and by no means concise) glossary but without losing the aim of the book, which wasn't to emulate past academic works on these topics. These works I have detailed more fully in the list of suggested further reading on page 237.

Appendix II: Timeline of major global events during the Age of Revolution

1756–63	The Seven Years' War (or French and Indian War).
1763	The Treaty (or Peace) of Paris.
1775–82	The American War of Independence.
1789	Destruction of the Bastille, commencement of the French Revolution.
1792–99	French Revolutionary Wars between republican France and most other European powers.
1793	Execution of Louis XVI and Jacobin dictatorship established to rule France.
Oct. 1797	Battle of Camperdown.
Feb. 1798–Aug. 1799	French expedition to Syria and Egypt.
Aug. 1799	Nelson destroys French naval support for French forces at the Battle of the Nile.
9 Nov. 1799	Napoléon Bonaparte appointed First Consul, effectively creating a military dictatorship.
Dec. 1799	Napoleonic Wars begin – France, with Spain and Holland to join as allies, against Britain, Austria, Prussia and Russia.
Aug. 1800	Battle of Copenhagen.
Mar. 1802–May 1803	Treaty (or Peace) of Amiens.
21 Oct. 1805	Nelson destroys French naval power in Europe at the Battle of Trafalgar.
3 Dec. 1810	Mauritius falls to a British invasion fleet.
11 Apr. 1814	Napoléon abdicates and is exiled to Elba.
Mar.– Jun. 1815	Napoléon returns for a hundred days.
18 Jun. 1815	Battle of Waterloo ends the Napoleonic Wars.

Appendix III: Navigators' timeline, before and during the Age of Revolution

1606	Willem Jansz visits the Gulf of Carpentaria (*Duyfken*).
1616	Dirk Hartog visits Shark Bay.
1629	Wreck of *Batavia*.
1642	Tasman discovers Van Diemen's Land and New Zealand (then Staaten Island); proves the two are separated.
1644	Tasman charts the north Australian coastline.
1656	Wreck of *Vergulde Draeke*.
1694	Disappearance of *Ridderschap van Hollandt*.
1697	De Vlamingh charts part of the west Australian coastline looking for survivors of *Ridderschap van Hollandt*; replaces Hartog's pewter plate at Shark Bay.
1688, 1699	Dampier makes two brief visits to the north-west coastline.
1763	Bougainville settles colony in the Falklands.
1764–65	Byron's voyage.
1766–68	Wallis's voyage.
1766–69	Carteret's voyage.
1767–69	Bougainville's voyage into the Pacific.
1768–70	Surville's voyage.
1768–71	Cook's first voyage (*Endeavour*).
1771–73	Marion's voyage (also known as Crozet's voyage).
1772–73	Kerguelen's and St Allouarn's voyages.
1772–75	Cook's second voyage (*Resolution*, *Adventure*).
1776–80	Cook's third voyage (*Resolution*, *Discovery*).
1778	Joseph Banks elected president of the Royal Society.
1785–88	La Pérouse's voyage.
May 1787–Jan. 1788	First Fleet sails to Botany Bay.

1787–1790	Bligh's first breadfruit voyage (in *Bounty* and *Bounty's* launch).
1790–91	Hunter returns to Britain in *Waaksamheyd*.
1790–92	Voyage of the *Pandora*.
1791–93	D'Entrecasteaux's voyage.
1792–93	Bligh's second breadfruit voyage (*Providence* and *Assistant*).
1794	Battle of 'Glorious First of June'.
Jan.–Sept. 1795	Bass, Flinders, Hunter, Waterhouse sail to Port Jackson on *Reliance* with *Supply*.
26 Oct.–3 Nov. 1795	Bass, Flinders, Martin sail to Georges River in *Tom Thumb* #1.
Jan.–Mar. 1796	Bass, Flinders sail to Norfolk Island on *Reliance*.
25 Mar.–2 Apr. 1796	Bass, Flinders, Martin sail to Lake Illawarra on *Tom Thumb* #2.
1796–98	Baudin's voyage to Tenerife, Puerto Rico and France.
Jun. 1796	Bass's fifteen-day attempt to cross the Blue Mountains.
Sept. 1796–Jun. 1797	Bass, Flinders voyage to Cape Town for cattle and sheep on *Reliance* and return.
Nov. 1796–Feb. 1797	*Sydney Cove* leaves Bengal and is wrecked in the Furneaux Islands off Van Diemen's Land.
17 May 1797	*Sydney Cove* survivors rescued by a fishing party at Wattamolla.
5–13 Aug. 1797	Bass finds coal with two *Sydney Cove* survivors in Hunter's whaleboat.
3 Dec. 1797–25 Feb. 1798	Bass discovers the Shoalhaven, Twofold Bay, Western Port and Bass Strait in Hunter's whaleboat with John Thistle and others.
1 Feb.–9 Mar. 1798	Flinders surveys NSW south coast on *Francis*.
May–Jul. 1798	Bass, Flinders, Thistle voyage to Norfolk Island on *Reliance*.

7 Oct. 1798– 12 Jan. 1799	Bass, Flinders, Thistle sail through Bass Strait, around Van Diemen's Land and prove no connection with the mainland in *Norfolk*.
29 May 1799– 4 Aug. 1800	Bass takes sick leave and departs for London via Macau on *Nautilus*.
Jul.–Aug. 1799	Flinders, Bongaree and Thistle head north to Moreton Bay on *Norfolk*.
Feb.–Aug. 1800	Flinders brothers and Waterhouse return to Britain on *Reliance*.
18 Mar.–16 Dec. 1800	Grant sails from Portsmouth to Port Jackson, first to coast the Victorian shores of Bass Strait, on *Lady Nelson*.
Jun. 1800	The French Government applies to Banks for a pass port for Baudin's expedition (*Géographe* and *Naturaliste*).
6 Sept. 1800	Flinders writes 'bold dash' letter to Banks proposing the *Investigator* expedition.
8 Oct. 1800	Bass married in London to Waterhouse's sister, Elizabeth.
19 Oct. 1800	Baudin's expedition leaves France for Australia (*Géographe* and *Naturaliste*).
Jan.–Aug. 1801	Bass sails from Britain to Port Jackson (*Venus*).
Mar.–May 1801	Grant, Barrallier and Caley return to chart Western Port (*Lady Nelson*).
May 1801	Baudin reaches Western Australia, charts from Geographe Bay north to Dampier Archipelago before retiring to Timor for refit (*Géographe* and *Naturaliste*).
18 Jul.–6 Dec. 1801	Flinders sails from Britain to Western Australia, begins to chart southern coastline (*Investigator*).

Sept. 1801	Grant resigns and returns to Britain in November 1801.
21 Nov. 1801– 14 Nov. 1802	Bass's first pork voyage to Tahiti and Hawaii and return to Port Jackson (*Venus*).
Nov. 1801–Feb. 1802	Baudin sails from Timor to Van Diemen's Land (*Géographe* and *Naturaliste*).
Nov. 1801–Feb. 1802	Murray discovers Port Phillip, charts King Island (*Lady Nelson*).
22 Feb. 1802	Thistle, Taylor and six others drown at Port Lincoln (*Investigator*).
Feb.–Mar. 1802	Flinders charts Denial Bay, north coast of Kangaroo Island, Spencer Gulf and Gulf St Vincent (*Investigator*).
6–9 Mar. 1802	Hydrographer Boullanger and seven others lost in dinghy off Freycinet Pen. *Géographe* and *Naturaliste* separate.
Mar. 1802	Baudin searches for Boullanger then coasts from Wilsons Promontory west along the southern coast line (*Géographe*). Hamelin sails to Port Jackson (*Naturaliste*).
8 Apr. 1802	Flinders and Baudin meet at Encounter Bay (*Géographe* and *Investigator*).
9 Apr.–20 Jun. 1802	Baudin examines Denial Bay, Kangaroo Island, Spencer Gulf and Gulf St Vincent, returns to Van Diemen's Land and thence to Port Jackson (*Géographe*).
22 Jul.–26 Nov. 1802	Flinders heads to chart the regions around Gladstone, Rockhampton and Mackay (*Investigator* and *Lady Nelson*) then to Sweers Island in the Gulf of Carpentaria (*Investigator*).
Nov. 1802–Apr. 1803	Baudin voyages to King Island (*Géographe*, *Naturaliste* and *Casuarina*), the mainland south coast

and the mainland north-west coast (for the second time) before retiring to Timor (in April) for water (*Géographe* and *Casuarina*). *Naturaliste* returns to France via Mauritius.

Dec. 1802–Jun. 1803	Flinders nurses rotten *Investigator* back to Port Jackson via Timor in March 1803.
Jun.–Jul. 1803	Baudin leaves Timor for a third visit to the mainland north-west coast, terminates survey and sails for Mauritius (*Géographe* and *Casuarina*).
7 Aug. 1803	Baudin reaches Mauritius (*Géographe* and *Casuarina*) and dies six weeks later, aged forty-nine.
17 Aug. 1803	Flinders wrecked on Wreck Reef (*Porpoise* and *Cato*).
26 Aug.–8 Sept. 1803	Flinders sails back to Port Jackson from Wreck Reef in the cutter *Hope*.
21 Sept.–7 Oct. 1803	Flinders returns to Wreck Reef (*Cumberland, Rolla, Francis*).
11 Oct.–16 Dec. 1803	Flinders sails from Wreck Reef through Torres Strait – stopping at Timor for five days in November – to Mauritius; arrives at Port Nord-Oest (*Cumberland*).
16 Dec. 1803	Baudin's expedition (now under Lieutenant Milius) departs Port Louis for France (*Géographe*).
17 Dec. 1803	Flinders arrives at Port Louis, Mauritius (*Cumberland*).
Jul. 1807	Péron's *Voyage* published.
Jul. 1810	Flinders released from Mauritius.
14 Dec. 1810	Péron dies aged thirty-five.
1811	Péron's (actually Freycinet did the work following Péron's demise) *Atlas, Deuxième Partie* published, containing the first complete map of Australia ever published and thirteen other folding maps and plans.

1812	Freycinet's navigational and geographic *Atlas* published, of thirty-two charts.
18 Jul. 1814	Flinders' *Narrative* (with atlas of charts) published.
19 Jul. 1814	Flinders dies aged forty.
1815	Freycinet completes and publishes volume 2 of Péron's *Voyage*, containing navigational and geographical narrative (mostly written by Péron), plus the atlas from 1812 issue, and for the first time mentioning Baudin by name.
7 Dec. 1818	Vice-Admiral Bligh dies aged sixty-four.
19 Jun. 1820	Sir Joseph Banks dies aged seventy-seven.

ENDNOTES

CHAPTER THREE

1 Letter to John Walker, 17 August 1771, in J. C. Beaglehole, *The Life of Captain James Cook*, Stanford University Press, California, 1974, p. 276.

CHAPTER FOUR

1 H. B. Carter, 1988, *Sir Joseph Banks*, British Museum (Natural History), London, p. 95.

2 John Gascoigne, 1994, *Joseph Banks and the English Enlightenment: Useful Knowledge and Polite Culture,* Press Syndicate of the University of Cambridge, Cambridge, UK, p. 3 and generally ch. 5.

3 David Mackay, 1985, *In the Wake of Cook: Exploration, Science & Empire 1780–1801*, Victoria University Press, Wellington, p. 176.

CHAPTER SIX

1 See generally, John Gascoigne, 1998, *Science in Service of the Empire: Joseph Banks, the British State and the Uses of Science in the Age of Revolution*, Press Syndicate of the University of Cambridge, Cambridge, UK, and also John Gascoigne, 1994, *Joseph Banks and the English Enlightenment: Useful Knowledge and Polite Culture,* Press Syndicate of the University of Cambridge, pp 5, 112–13, 167–68.

2 Banks, 28 April 1801, in F. M. Bladen, 1896, *Historical Records of New South Wales*, NSW Government Printer, Sydney, vol. iv, p. 348. Although not signed by Nepean, has generally been accepted as written by him.

CHAPTER SEVEN

1 Ray Parkin, 1997, *HM Bark Endeavour*, Melbourne University Press, Carlton, Victoria, p. 43.

2 M. Lewis, 1966, in Greg Dening, 1992, *Mr Bligh's Bad Language: Passion, Power and Theatre on the Bounty*, Press Syndicate

of the University of Cambridge, Cambridge, UK, p. 120.

3 Ray Parkin, op. cit.

4 J. C. Beaglehole, 1974, *The Life of Captain James Cook*, Stanford University Press, California, p. 610.

5 ibid., pp 619, 620 footnote 1.

6 ibid., p. 641.

7 ibid., pp 559–60.

8 ibid., p. 560.

CHAPTER EIGHT

1 Greenwich National Maritime Museum, Flinders Papers 60/017.FLI/1 in Geoffrey Ingleton, 1986, *Matthew Flinders, Navigator and Chartmaker*, Genesis Publications, Surrey, UK and Hedley, Victoria, Australia, p. 17.

2 George Mackaness, 1931, *The Life of Admiral Bligh*, Angus & Robertson, Sydney, vol. II, p. 64.

CHAPTER NINE

1 Geoffrey Ingleton, 1986, *Matthew Flinders, Navigator and Chartmaker*, Genesis Publications, Surrey, UK and Hedley, Victoria, p. 9.

2 ibid., p. 10.

3 ibid., p. 24.

4 Governor Hunter to the Duke of Portland, 1 March 1798, as quoted in Keith Bowden, 1952, *George Bass 1770–1803*, Oxford University Press, London, p. 58.

5 Ernest Scott, 1914, *The Life of Matthew Flinders*, Angus & Robertson, Sydney, p. 83.

6 Letter from Flinders to Bass on exhibition by the State Library of NSW, viewed by the author in October 2001, bearing ref. no.

ZML MSS 7046.

7 Ernest Scott, loc. cit.

8 Keith Bowden, op. cit., pp 60, 74.

9 David Collins, 1802, *An Account of the English Colony in New South Wales,* vol. ii, Cadell & Davies, London, p. 85.

10 Ingleton, op. cit., p. 36.

11 Keith Bowden, op. cit., p. 97.

12 Bladen, 1897, *Historical Records of New South Wales*, NSW Government Printer, vol. v, p. 15.

13 ibid.

14 ibid.

15 ibid.

16 Keith Bowden, op. cit., p. 137.

CHAPTER TEN

1 Letter reported in *The Australian*, 21 September 2001.

2 Geoffrey Ingleton, 1986, *Matthew Flinders, Navigator and Chartmaker*, Genesis Publications, Surrey, UK, and Hedley, Victoria, p. 50.

3 ibid.

4 ibid., p. 91, citing Greenwich National Maritime Museum, Flinders Papers 60/175.FL1/25.

CHAPTER ELEVEN

1 Ida Lee, 1911, *The Logbooks of the Lady Nelson*, Grafton & Co., London, p. 317.

2 F. M. Bladen, 1897, *Historical Records of New South Wales*, NSW Government Printer, Sydney, vol. v, p. 460.

CHAPTER TWELVE

1 F. B. Horner, 1995, *Looking for La Pérouse*, Melbourne University Press, Carlton, Victoria, p. 7.

CHAPTER THIRTEEN

1 F. B. Horner, 1995, *Looking for La Pérouse*, Melbourne University Press, Carlton, Victoria, p. 51.

CHAPTER FOURTEEN

1 F. B. Horner, 1987, *The French Reconnaissance: Baudin in Australia*, Melbourne University Press, Carlton, Victoria, p. 106.
2 Nicolas Baudin, *The Journal of Post Captain Baudin*, trans. Patricia Cornell, 1974, Adelaide Libraries Board of South Australia, Adelaide, p. 355.

CHAPTER FIFTEEN

1 Greenwich National Maritime Museum, Flinders Papers 60/017.FLI/25. Also quoted in Geoffrey Ingleton, 1986, *Matthew Flinders, Navigator and Chartmaker*, Genesis Publications, Surrey, UK and Hedley, Victoria, p. 104.
2 ibid.
3 F. M. Bladen, 1896, *Historical Records of New South Wales*, NSW Government Printer, Sydney, vol. iv, p. 372.
4 Ingleton, op. cit., pp 107–08.
5 ibid., p. 110.
6 Letter from Flinders to Bass on exhibition by the State Library of NSW, viewed by the author in October 2001, bearing ref. no. ZML MSS 7046.
7 J. C. Beaglehole, 1974, *The Life of Captain James Cook*, Stanford University Press, California, p. 307.
8 Matthew Flinders, 1814, *A Voyage to Terra Australis*, G & W Nicol, London, vol. 1, pp 109, 113.

9 ibid., p. 117.

CHAPTER SIXTEEN

1 Matthew Flinders, 1814, *A Voyage to Terra Australis*, G & W Nicol, London, vol. 1, p. 193.
2 Nicolas Baudin, *The Journal of Post Captain Baudin*, trans. Patricia Cornell, 1974, Adelaide Libraries Board of South Australia, Adelaide, p. 383.
3 ibid., p. 401.
4 ibid., p. 393.
5 ibid., p. 420.
6 Flinders, op. cit., p. 117.
7 ibid., p. 236.

CHAPTER SEVENTEEN

1 Matthew Flinders, 1814, *A Voyage to Terra Australis*, G & W Nicol, London, vol. 2, pp 83, 95.
2 Geoffrey Ingleton, 1986, *Matthew Flinders, Navigator and Chartmaker*, Genesis Publications, Surrey, UK and Hedley, Victoria, p. 193.
3 Flinders, op. cit., vol. 2, p. 104.
4 Ingleton, op. cit., pp 202, 204.
5 Flinders, op. cit., vol. 2, p. 123.
6 ibid., p. 132.
7 ibid., p. 147.
8 ibid.

CHAPTER EIGHTEEN

1 Nicolas Baudin, *The Journal of Post Captain Baudin*, trans. Patricia Cornell, 1974, Adelaide Libraries Board of South Australia, Adelaide, p. 434.
2 ibid., p. 441.
3 Hamelin's journal, quoted in F. B. Horner,

1987, *The French Reconnaissance: Baudin in Australia*, Melbourne University Press, Carlton, Victoria, p. 264.

4 Baudin, trans. Patricia Cornell, op. cit., p. 442.

5 F. M. Bladen, 1897, *Historical Records of New South Wales*, NSW Government Printer, Sydney, vol. v, appendix B, pp 831–32.

6 See Geoffrey Ingleton, 1952, *True Patriots All*, Angus & Robertson, Sydney, on the rumoured lawlessness on Bass Strait islands by the 1830s in relation to the loss of the ship *Britomart*.

7 F. B. Horner, 1987, *The French Reconnaissance: Baudin in Australia*, Melbourne University Press, Carlton, Victoria, p. 269, quoting Charles Bateson, 1973, *Dire Strait: A History of Bass Strait*, Sydney, pp 18–20.

8 Baudin, op. cit., p. 467.

9 ibid., p. 470.

10 ibid.

11 ibid., p. 480.

12 ibid., p. 498.

13 ibid.

14 ibid., p. 510.

15 ibid., p. 525.

16 ibid., p. 537.

17 ibid., p. 521.

18 ibid., p. 544.

19 ibid., p. 557.

20 ibid., p. 560.

CHAPTER NINETEEN

1 Matthew Flinders, 1814, *A Voyage to Terra Australis*, G&W Nicol, London, vol. 2, p. 173.

2 ibid., p. 183.

3 ibid., p. 188.

4 Phillis Edwards, 1981, *The Journal of Peter Good*, British Museum of Natural History, London, p. 111.

5 ibid.

6 Flinders, op. cit., p. 198.

7 ibid., p. 197.

8 ibid., p. 205.

9 ibid., p. 209.

10 ibid., pp 198, 212.

11 ibid., p. 232 note.

12 ibid., p. 228.

13 ibid., p. 232.

14 ibid., p. 248.

15 ibid., p. 271.

CHAPTER TWENTY

1 Matthew Flinders, 1814, *A Voyage to Terra Australis*, G & W Nicol, London, vol. 2, p. 296.

2 Brown to Banks, 6 August 1803, *HRNSW*, vol. v, p. 181.

3 F. B. Horner, 1987, *The French Reconnaissance: Baudin in Australia*, Melbourne University Press, Carlton, Victoria, p. 357.

4 Flinders, op. cit., p. 314.

5 *Historical Records of Australia*, vol. iv, p. 399, King to Nepean 17 Sept. 1803.

6 Flinders, op. cit., p. 323.

7 ibid., p. 326.

8 ML MSS Safe 1/54 Flinders public letters 1803–07. Cited in Geoffrey Ingleton, 1986, *Matthew Flinders, Navigator and Chartmaker*, Genesis Publications, Surrey, UK and Hedley, Victoria, p. 246.

9 Flinders, op. cit., p. 328.

10 ibid.

11 ibid., p. 347.

12 ibid., p. 352.

13 ibid., p. 354.

CHAPTER TWENTY-ONE

1 Ernest Scott, 1914, *The Life of Matthew Flinders,* Angus & Robertson, Sydney, p. 439.

2 ibid., p. 454.

3 ibid., p. 459.

CHAPTER TWENTY-TWO

1 Matthew Flinders, 1814, *A Voyage to Terra Australis*, G & W Nicol, London, vol. 2, p. 360.

2 ibid., p. 363.

3 ibid, p. 364.

4 Geoffrey Ingleton, 1986, *Matthew Flinders, Navigator and Chartmaker*, Genesis Publications, Surrey, UK and Hedley, Victoria, p. 271.

CHAPTER TWENTY-THREE

1 Horner, 1987, *The French Reconnaissance: Baudin in Australia*, Melbourne University Press, Carlton, Victoria, p. 82.

2 Audiat: F. Péron, p. 62; Horner, op. cit., p. 339.

CHAPTER TWENTY-FOUR

1 Ernest Scott, 1914, *The Life of Matthew Flinders,* Angus & Robertson, Sydney, p. 390.

2 Geoffrey Ingleton, 1986, *Matthew Flinders, Navigator and Chartmaker*, Genesis Publications, Surrey, UK and Hedley, Victoria, p. 297; although Baudin recorded in his journal the name as 'Golfe de la Mauvaise' for the suffering it caused his crew (see ch. 16).

CHAPTER TWENTY-FIVE

1 Matthew Flinders, 1814, *A Voyage to Terra Australis*, G & W Nicol, London, vol. 1, p. iii.

2 François Péron, 1809, *A Voyage to the Southern Hemisphere*, trans. Richard Philips, Blackfriars, p. 47 (reprinted by Marsh Walsh Publishing, Melbourne, 1975).

3 *Victorian Geographical Journal* incl. proceeds of the Royal Geographical Society of Australasia, vol. XXVIII, 1910–11, in Geoffrey Ingleton, 1986, Matthew Flinders, Navigator and Chartmaker, Genesis Publications, Surrey, UK, and Hedley, Victoria, p. 420.

GLOSSARY

ABLE SEAMAN Seaman promoted from ordinary seaman, the lowest adult rank.

ANTHROPOLOGY The study of the origins, physical development, customs and beliefs of man from a scientific perspective, rather than a theological base.

BOATSWAIN OR BOSUN Warrant (i.e. non-commissioned) officer responsible for maintenance of the ship's rigging, anchors and cables.

BOWSPRIT Spar protruding forward from a sailing ship's bow, from which a spritsail and spritsail topsail were set (from yards) and to which the tacks, or leading lower corners, of triangular jib sails were attached.

CHRONOMETER Timepiece invented by John Harrison, used to calculate a ship's longitude. It recorded time at Greenwich for comparison with local time – the difference indicating, relative to the sun at noon, how far from the Greenwich meridian of longitude a ship was. It was regularly wound and rated to ensure it didn't gain or lose time at more than a known rate. The only alternative means of calculating this time was by taking astronomical observations from a stable platform and calculating time from these with complicated tables.

COMMANDER Next rank above lieutenant in the British Navy. Gained by service and recommendation. Essential qualification for command of ships (over a certain tonnage). There was initially some debate as to whether *Investigator* was too small to require a commander, rather than a first lieutenant, to command her. Cook was promoted from master only to lieutenant just before (and for the purposes of) the voyage on *Endeavour* in 1768, and *Endeavour* was only slightly smaller than *Investigator*.

HYDROGRAPHY The study of the ocean's surface waters, particularly for navigation.

LIEUTENANT Lowest rank of commissioned officer in the British Navy. A commission was gained by passing a lieutenant's exam, which required a minimum of six years of service at sea as a prerequisite.

LONGBOAT OR LAUNCH Usually ten-oared boat used as a ship's general-purpose heavy work boat (capable of carrying a large anchor or many water barrels); over 18 feet in length (*Bounty*'s longboat was 23 feet long) but very broad in beam (6 feet or more) and of slightly deeper stern draught than the pinnace (3 feet 2 inches). The longboat could step two masts and sails, often with a third small lugsail over the stern.

MARINE Seaborne contingent of soldiers, the number of which varied with the size of the vessel.

MASTER Most senior non-commissioned officer, or warrant officer, in the British Navy at the time. A master was responsible for the navigation of a ship, subject to the command of its officers. It was an archaic

rank from the days when the officers were soldiers who knew nothing of sailing and required sailors, subject to their commands, to sail the ship. It wasn't unusual in smaller ships for the senior officer to have the rank of master and commander. As commander of *Investigator*, Flinders was assisted by both a first lieutenant (Fowler) and a master (Thistle, then Aken), probably because Flinders had the additional burden of being the only capable hydrographer on board.

MATE Assistant warrant officer to a senior warrant officer. Hence bosun's mate, master's mate and so forth.

PAROLE A promise given by one in captivity, as 'an officer and a gentleman' – usually to not attempt to escape or take up arms – in order to receive some benefit from the captor. The usual benefit was either temporary or permanent release or return of one's sword.

PASSPORT A written document from a State ordering uninterrupted passage and support be given to the specified bearer – usually a scientific expedition from an enemy State – provided stipulated conditions were adhered to – typically that no trade, espionage, carrying of dispatches or act of war was engaged in.

PINNACE Usually six-oared boat – often called a barge – used by officers or for official purposes; at approximately 18 feet in length, almost as long as the longboat and much narrower and sleeker (approx. 5-foot beam) and slightly shallower draught (approx. 3 feet). The pinnace could step two masts and sails.

POST-CAPTAIN The next rank above commander in the Royal Navy, and a necessary rank to obtain in order to be eligible to command a frigate or ship of the line, and in order to obtain flag rank as a rear-admiral. A post-captain's name was entered at the foot of a post-captains' list upon promotion, and seniority on this list was one determinant of one's eligibility for further promotion. Exemplary service could accelerate this process, but otherwise a post-captain, provided he lived long enough, could still expect promotion to rear-admiral as others above him on the post-captain's list were removed from it, from any of promotion, retirement, court martial or death. This wasn't a fast process, even in wartime. Bligh waited twenty-one years for this, from promotion to post-captain in 1790 to promotion to Rear-Admiral of the Blue in 1811. Robert Fowler, promoted post-captain in 1811, had to wait thirty-five years for promotion to rear-admiral during peacetime.

REAR-ADMIRAL, VICE-ADMIRAL, ADMIRAL Three ranks of admiral in the British Navy, and for all three there were three classes based in flag colour (again in ascending order), blue, white and red. This meant theoretically there could be no more than nine admirals in the British fleet, which was the case when the system was initially imposed in the 1660s but which was out of date by the time the fleet had expanded in the 1800s to many separate fleets.

SEXTANT Instrument used to determine angular distances, usually of celestial bodies from the horizon, in order to calculate a ship's latitude and to assist in calculating longitude.

SHIP OF THE LINE A naval ship with more than sixty-four guns, which was a measure of her total broadside firepower (a broadside was the effect of all guns on one side, i.e. thirty-two or more, going off together). Naval tactics of the day required opposing fleets, each fleet in an extended line, one ship behind the other, to engage in conflict by sailing those extended lines past each other in opposing directions, exchanging gunfire. As only one side of a ship's armament was therefore used at any time, the broadside force of its guns was of great importance.

SOUNDING A method of measuring water depth in shoaling or shallow water by use of a lead weight on the end of a marked line of thin strong twine. The lead weight had tallow or wax in a small hollow on its tip. It took some skill to cast the lead cleanly from a position low on a ship's bow in order that it landed sufficiently far away that it could sink to the sea bottom by the time the ship passed above the impact point for an accurate reading. When hauled up, the nature of the bottom could be ascertained from what was stuck to the weight's tallow or wax, which was important for anchoring purposes.

WEAR SHIP A manoeuvre for changing a

ship's direction while sailing into the wind,
by turning away from the wind, continuing
in a circle of approximately 270 degrees,
with the ship's stern passing across the
direction from which the wind blows, until
meeting the new desired windward course.
This differs to tacking, which achieves the
same result, but by which the ship's bow,
rather than its stern, passes across the wind.

YARD Horizontal spars slung from their centre to masts and from which sails were sus-
pended.

Suggested further reading

ON PEOPLE AND SHIPS

Joseph Banks

Beaglehole, J. C. 1962, *The Endeavour Journal of Joseph Banks 1768–1771*, The Trustees of the Public Library of NSW in association with Angus & Robertson, Sydney.

Bladen, F. M. 1896–97, *The Historical Records of New South Wales*, vols iv & v, NSW Government Printer, Sydney.

Gascoigne, John 1994, *Joseph Banks and the English Enlightenment: Useful Knowledge and Polite Culture,* Press Syndicate of the University of Cambridge, Cambridge.

Carter, H. B. 1988, *Sir Joseph Banks*, British Museum (Natural History), London.

Mackaness, George 1936, *Sir Joseph Banks: His relations with Australia*, Angus & Robertson, Sydney.

O'Brian, Patrick 1997, *Joseph Banks*, Harvill Press, London.

Matthew Flinders

Baker, Sidney 1962, *My Own Destroyer: A Biography of Matthew Flinders, Explorer and Navigator*, Currawong Publishing Co., Sydney.

Bladen, F. M. 1896–97, *The Historical Records of New South Wales*, vols iv & v, NSW Government Printer, Sydney.

Edwards, Phillis 1981, *The Journal of Peter Good*, British Museum of Natural History, London.

Flinders, Matthew 1814, *A Voyage to Terra Australis*, G & W Nicol, London.

Ingleton, Geoffrey 1986, *Matthew Flinders, Navigator and Chartmaker*, Genesis

Publications, Surrey, UK & Hedley,
Victoria.

Mack, James D. 1966, *Matthew Flinders*,
Thomas Nelson, Melbourne.

Scott, Ernest 1914, *The Life of Matthew
Flinders*, Angus & Robertson, Sydney.

Nicolas Baudin

Baudin, Nicolas *The Journal of Post Captain
Baudin*, trans. Patricia Cornell 1974,
Adelaide Libraries Board of South
Australia, Adelaide.

Bladen, F. M. 1896–97, *The Historical
Records of New South Wales*, vols iv & v,
NSW Government Printer, Sydney.

Dunmore, John 1997, *Visions & Realities:
France in the Pacific 1695–1995*,
Heritage Press, Waikanae, New Zealand.

Flinders, Matthew 1814, *A Voyage to Terra
Australis*, G & W Nicol, London.

Horner, F. B. 1987, *The French
Reconnaissance: Baudin in Australia*,
Melbourne University Press, Carlton,
Victoria.

Péron, François 1809, *A Voyage to the
Southern Ocean*, vol. 1, Richard Phillips,
London.

Scott, Ernest 1914, *The Life of Matthew
Flinders*, Angus & Robertson, Sydney.

François Péron

Horner, F. B. 1987, *The French
Reconnaissance: Baudin in Australia*,
Melbourne University Press, Carlton,
Victoria.

Ingleton, Geoffrey 1986, *Matthew Flinders,
Navigator and Chartmaker*, Genesis

Publications, Surrey, England & Hedley,
Victoria.

Péron, François 1809, *A Voyage to the
Southern Ocean*, vol. 1, Richard Phillips,
London.

Scott, Ernest 1914, *The Life of Matthew
Flinders*, Angus & Robertson, Sydney.

Wallace, Colin 1984, *The Lost Australia of
François Péron*, Nottingham Court Press,
London.

George Bass

Bladen, F. M. 1896–97, *The Historical
Records of New South Wales*, vols iv & v,
NSW Government Printer, Sydney.

Bowden, Keith 1952, *George Bass
1770–1803*, Oxford University Press,
Amen House, London.

Collins, David 1798, 1802, *An Account of
the English Colony of New South Wales*,
vols i & ii, T. Cadell & W. Davies,
London, Australian edn published in
1975 by AH & AW Reed, Sydney.

Flinders, Matthew 1814, *A Voyage to Terra
Australis*, G & W Nicol, London.

Ingleton, Geoffrey 1986, *Matthew Flinders,
Navigator and Chartmaker*, Genesis
Publications, Surrey, UK & Hedley,
Victoria.

Scott, Ernest 1914, *The Life of Matthew
Flinders*, Angus & Robertson, Sydney.

William Bligh

Dening, Greg 1992, *Mr Bligh's Bad
Language: Passion, Power and Theatre on
the Bounty*, Press Syndicate of the
University of Cambridge, Cambridge.

Hough, Richard 2000, *Captain Bligh and*

Mr Christian, Chatham Publishing, London.

Mackaness, George 1931, *The Life of Admiral Bligh*, Angus & Robertson, Sydney.

James Grant

Grant, James 1803, *Narrative of a Voyage of Discovery in the Lady Nelson*, T. Egerton, London.

Ingleton, Geoffrey 1986, *Matthew Flinders, Navigator and Chartmaker*, Genesis Publications, Surrey, England & Hedley, Victoria.

Lee, Ida 1911, *The Logbooks of the Lady Nelson*, Grafton & Co., London.

James Cook

Beaglehole, J. C. 1974, *The Life of Captain James Cook*, Stanford University Press, Stanford, California.

Hawkesworth, John 1773, *An Account of the Voyages for Making Discoveries in the Southern Hemisphere*, A. Leathley and others, Dublin.

Hough, Richard 1995, *Captain James Cook, a Biography*, Hodder & Stoughton, London.

Parkin, Ray 1997, *HM Bark Endeavour*, Melbourne University Press, Carlton, Victoria.

Earlier French navigators

Dunmore, John 1997, *Visions & Realities: France in the Pacific 1695–1995*, Heritage Press Ltd, Waikanae, New Zealand.

Dunmore, John 1965, *French Explorers in the Pacific*, vols 1 & 2, Oxford University Press, London.

Horner, F. B. 1995, *Looking for La Pérouse*, Melbourne University Press, Carlton, Victoria.

La Pérouse, 1798, *The Voyage of La Pérouse Round the World in the Years 1785, 1786, 1787, 1788*, J. Stockdale, London.

Ling Roth, H. 1891, *Crozet's Voyage to Tasmania*, Truslove & Shirley, London.

Scott, Ernest 1912, *Lapérouse*, Angus & Robertson, Sydney.

Earlier British navigators

Badger, Geoffrey 1988, *The Explorers of the Pacific*, Kangaroo Press, 2nd edn, Kenthurst, NSW.

Flinders, Matthew 1814, *A Voyage to Terra Australis*, G & W Nicol, London.

Hawkesworth, John 1773, *An Account of the Voyages for Making Discoveries in the Southern Hemisphere*, A. Leathley and others, Dublin.

Endeavour

Parkin, Ray 1997, *HM Bark Endeavour*, Melbourne University Press, Carlton, Victoria.

Lady Nelson

Grant, James 1803, *Narrative of a Voyage of Discovery in the Lady Nelson*, T. Egerton, London.

Lee, Ida 1915, *The Logbooks of the Lady Nelson*, Grafton & Co., London.

ON EVENTS, LIFE AND POLITICS

Enlightenment and natural history

Badger, Geoffrey 1988, *The Explorers of the Pacific*, Kangaroo Press, 2nd edn, Kenthurst, NSW.

Finney, C. M. 1984, *To Sail Beyond the Sunset: Natural History in Australia 1699–1801*, Rigby Publishers, Adelaide.

Gascoigne, John 1994, *Joseph Banks and the English Enlightenment: Useful Knowledge and Polite Culture,* Press Syndicate of the University of Cambridge, Cambridge, UK.

Gascoigne, John 1998, *Science in Service of the Empire: Joseph Banks, the British State and the Uses of Science in the Age of Revolution,* Press Syndicate of the University of Cambridge, Cambridge, UK.

Mackay, David 1985, *In the Wake of Cook: Exploration, Science & Empire, 1780–1801,* Victoria University Press, Wellington.

Longitude

Badger, Geoffrey 1988, *The Explorers of the Pacific*, Kangaroo Press, 2nd edn, Kenthurst, NSW.

Sobel, Dava and Andrewes, William 1998, *The Illustrated Longitude*, Fourth Estate, London.

The Seven Years' War

Anderson, Fred 2001, *Crucible of War*, Vintage Books, Random House, New York.

Hydrography

Blewitt, Mary 1957, *Surveys of the Seas: A Brief History of British Hydrography*, May Gibbon & Key, London.

Edgell, Vice Admiral Sir John 1965, *Sea Surveys: Britain's Contribution to Hydrography*, HM Stationary Office, London.

Ritchie, Rear-Admiral G. S. 1968, *The Admiralty Chart, British Naval Hydrography in the Nineteenth Century,* Hollis & Carter, London.

Index of people, places and ships

Adventure, 27, 100
Ahu-toru, 9, 92–3
Aken, John, 141, 189
Allen, John, 83, 176
American War of Independence, 31–2
Atlas, Deuxième Partie (Freycinet), 197–203
Atlas Historique (Péron), 195–6
Australia and Australians, terms first used, 166, 209

Banks, Joseph, 5, 17, 21, 25, 26–7, 45–6, 217–18
  advisory role to the state, 33–5, 62–3
  disappointment with Flinders, 123–4, 135
  Enlightenment, and, 22–3
  importance of breadfruit, 24
  patron of Bligh, 54
  patron of Flinders, 80–2

  plant transhipment, 23–4
Barrallier, Francis, 87
Barrier Reef, naming of, 144
Bass, George, 68–77
Bass Strait exploration, 72–3
Batavia (Jakarta), 48
Batchelor's Delight, 12
Baudin, Charles, 151, 162
Baudin, Nicolas, 5, 80, 106–22, 213
  encounter with Flinders, 130–3
  exploration of Van Diemen's Land, 136–8
  at King Island, 150–5
  stay at Port Jackson, 138–42
  in Western Australia and Timor, 156–62
Bauer, Ferdinand, 83, 171, 212
Bligh, William, 5, 28, 49, 53–63, 65, 216–17
  breadfruit and *Bounty*, 24, 49, 55–6
  breadfruit and *Providence*, 28, 57–9, 64–5

Bongaree, 77, 141
Bougainville, Louis Antoine de, 2, 6, 8–10, 12
Boullanger, Charles-Pierre, 119–20
Bounty mutiny, 49, 55–9
Bridgewater, 171–3, 178
Brown, Robert, 83, 169, 170–1, 211–12
Byron, John, 10–11

Caley, George, 87
Carteret, Philip, 12–13
Casuarina, 140, 150–1, 158–62
Cato, 171
Chatham, 29
Clerke, Charles, 13, 17–18, 29
Collins, David, 134–5
Cook, James, 2, 5, 13–14, 17–19
    charting of New Zealand, 18
    development of future navigators, 28–30
    first voyage, 15, 17–20
    leadership style, 50–2
    scurvy prevention, 46
    second voyage, 26–7
    third voyage, 27–8, 32, 54
Crosley, John, 83, 126, 208
Cumberland, 150–1, 174–6, 178–9

Dalrymple, Alexander, 16, 83–4, 189
Dampier, William, 2, 12
Davis, Edward, 12
Decaen, Charles Mathieu Isadore, 181–2
d'Entrecasteaux, Bruny, 47, 98–102
Dilba, 71
Director, 60
Discovery, 28, 29
Dolphin, 10–12
Dumont D'Urville, J, 216

Dutch explorers, 2
    accuracy of charts, 41, 146–8, 160–1, 166–7

Endeavour, 38
    crew of, 13, 17, 18
    importance of voyage, 15 18–20
    success of voyage, 18, 21
Espérance, 104

Falkland Islands, 7–8, 11
First Fleet, 34–6, 47
Flinders, Ann (née Chappelle), 82, 123–4, 204, 210
Flinders, Matthew, 5, 28, 57–60, 64–78, 79–84
    blunders, 123–9
    encounter with Baudin, 130–3
    encounters with Aborigines, 165–6
    imprisonment on Mauritius, 185–91
    magnetic distortion, and 208–9
    Port Jackson to Mauritius, 170–80
    return to England and publication of charts, 204–10
    stay at Port Jackson, 139
    voyage in Gulf of Carpentaria, 143–9, 163–8
Flinders, Samuel, 67, 77, 82, 146, 168, 175–7, 208
Fowler, Robert, 82, 170, 176–7
Francis, 174, 176
Franklin, John, 28, 82, 174, 176, 204, 212, 218
French explorers and exploration, 6–10, 36, 90–7, 215–16
Freycinet, Louis de, 43, 44, 158, 162, 195, 197–203, 215, 216

Furneaux, Tobias, 44, 100

Géographe, 44, 112, 120–2, 136–8, 155–62, 184
Gonneville Land, 92
Good, Peter, 83, 169, 211
Gore, John, 13, 17, 27
Grant, James, 86–8
Green, Charles, 17, 18
Gulf of Carpentaria exploration, 141, 143–9, 163–8

Haite, Francis, 17
Hamelin, Emanuel, 44, 115–16, 139, 151
Hamilton, Guy, 71
Harrington, 76
Hartog, Dirk, 116
Heywood, Peter, 58–9
Holland, Samuel, 13
Hope, 173
Hunter, John (anatomist), 69
Hunter, John (captain), 66, 68, 99

Illustrationes (Bauer), 212
Investigator, 82–4, 143–9, 163, 168–9

Kerguelen, Yves-Joseph de, 93–5
King, Phillip Gidley, 138–9, 152–3
King, Phillip Parker, 28, 40, 215, 218

La Pérouse, Jean-François Galaup, compte de, 95–9, 101–2
Labillardière, Jacques, 104, 212
Lady Nelson, 80, 85–9, 134, 140–1, 144–5
Leschenault, Jean-Baptiste, 211–12
Lesueur, Charles-Alexandre, 114, 195, 202, 212–13
Lind, James, 45

Marion du Fresne, Nicholas, 92–3
Matra, James, 35
Mauritius, 181–4
Menzies, Archibald, 30
midshipmen, 112
Molyneaux, Robert, 17
Monkhouse, Jonathan, 18
Murray, John, 87, 134, 145

Nanbaree, 141
Naturaliste, 44, 112, 115, 120–2, 139, 140, 150–1
Nautilus, 75
navigators
  skills, 4, 5
  crew morale, 48–9
  difficulties with midshipmen, 112
  crew health and scurvy, 44–8
  methods of navigation, 40–3
New Holland, 8
New South Wales penal colony, founding of 34–6
Norfolk, 73, 77, 128

Pandora, 58
Péron, François, 110–11, 113–14, 152, 171, 212–13
  publications, 192–202
  report to Decaen, 182–4
Petit, Nicolas Martin, 114, 195
Pickersgill, Dick, 17
Pitcairn Island, 12, 41
Pitt, Thomas, 30
Porpoise, 170–1
Port Phillip exploration, 134–5
Portuguese exploration, 16
Providence, 57–9, 65

Récherche, 104
Reliance, 66–8, 70–1, 75, 78
Resolution, 27, 28, 54
Rolla, 174, 176
Royal Society, 14, 15, 23
Rum Rebellion, 62

Seven Years War, 6–7
Sirius, 66–7
St Allouarn, François Alesno de, 93–4
South Australia exploration, 127–9
Supply, 67, 87
Surville, Charles de, 10, 91
Swallow, 12
Sydney Cove, 71

Tamar, 10
Tasman, Abel, 2, 8, 41
Tasmania exploration, *see* Van Diemen's
   Land
Taylor, William, 128
Terra Australis Incognita, 8, 12, 16, 27
Thistle, John, 71, 128
Timor, 48, 117–18
Tom Thumb, 69–70
Torres, L, 16

Treaty of Paris, 6–8
Treaty of Tordesillas, 7
Tuckey, James, 135

Vancouver, George, 27, 29–30, 216
Van Diemen's Land, exploration 73,
   99–101, 118–20
Vanikoro, Santa Cruz Islands, 101–2
Venus, 75
Venus, transit of, 15, 18
Victoria exploration, 88
Vlamingh, Willem de, 2
   pewter plate, 116
Voyage de Découvertes aux Terres Australes
   (Péron), 195–6

Waakzamheyd, 66–7
Walker, John, 19
Wallis, Samuel, 12
Weddell, James, 27
Wessel Islands, 168
Westall, William, 83, 165, 174, 176
Western Australia exploration, 100, 115–17
Whitby cats, 38
Wilkinson, Francis, 17
Wreck Reef, 172–6